Sustainable Slope Stabilisation using Recycled Plastic Pins

To our family members for their sacrifice, love and support

Sustainable Slope Stabilisation using Recycled Plastic Pins

Sahadat Hossain

The University of Texas at Arlington, Arlington, TX, USA

Sadik Khan

Jackson State University, Jackson, MS, USA

Golam Kibria

Arias Geoprofessionals, Inc., Arlington, TX, USA

CRC Press is an imprint of the
Taylor & Francis Group, an **informa** business

A BALKEMA BOOK

Applied for

Published by: CRC Press/Balkema
               P.O. Box 11320, 2301 EH Leiden, The Netherlands
               e-mail: Pub.NL@taylorandfrancis.com
               www.crcpress.com – www.taylorandfrancis.com

First issued in paperback 2020

ISBN 13: 978-0-367-57358-4 (pbk)
ISBN 13: 978-1-138-63610-1 (hbk)

Visit the Taylor & Francis Web site at
http://www.taylorandfrancis.com

and the CRC Press Web site at
http://www.crcpress.com

Typeset by MPS Limited, Chennai, India

Library of Congress Cataloging-in-Publication Data

Table of contents

Preface

The repair and maintenance due to highway slope failures costs millions of dollars each year in the United States. Highway slope failures are generally surficial in nature, characterised by a failure depth of less than three metres (ten feet). The possibility of highway slope failure is even higher for highway slopes built on expansive clayey soils. In expansive clayey soils, repeated moisture variation due to climatic change causes cracks along the top of highway slopes or highway shoulders. These cracks create pathways for rainwater to seep into the highway slopes. The seeped water increases pore water pressure, decreases soil strength, and eventually leads to highway slope failure.

The conventional slope stabilisation methods (drilled shaft, retaining wall, installation of soil nail, use of geogrids and others) can be expensive and time-consuming for the repair of shallow slope failure. Therefore, new, innovative, sustainable and cost-effective slope stabilisation methods are being tried, tested, and implemented increasingly in recent years. One such method is the use of recycled plastic pins (RPPs) for slope stabilisation.

Plastic and plastic products are non-degradable components of municipal solid waste (MSW). Globally, most of the generated plastics are discarded into landfill (90% in USA and 40% in Europe). Even though they are lightweight, they occupy large volumes of landfill space. As they are non-degradable, they remain in the landfill for a long period and keep occupying valuable landfill space, which otherwise could have been reused for more additional waste burial. Thus plastic, being non-degradable, is problematic if discarded into landfills; however, due to the same non-degradable nature, plastic can be very useful and beneficial for civil engineering infrastructure projects. Because soil slopes and highway soil slopes repaired with plastic products can retain their engineering characteristics for a long time, the use of RPPs, made out of recycled plastic bottles, for highway slope stabilisation and civil engineering infrastructure projects demonstrates a perfect example of sustainable engineering solutions.

RPPs are manufactured mainly using recycled soda bottles, and approximately 600 soda bottles are used for manufacturing one RPP. Increasing the use of RPPs for slope stabilisation could potentially divert millions of plastic bottles from landfill. Therefore, diverting or reusing plastic products from landfill could save landfill space in the short term, while helping to stabilise the landfill at a faster rate in the long run.

RPPs were successfully utilised for the first time for slope stabilisation in Missouri, USA in 1999. Since then many other highway slopes in Missouri, Kansas and, recently, in North Texas have been stabilised using RPPs. The University of Texas at Arlington (UTA), in collaboration with the Texas Department of Transportation's (TxDOT) Dallas District, successfully implemented slope stabilisation using RPPs for the first time in Texas in 2011. Since then, many slopes around the RPP-stabilised slope in Texas have failed over the last five years, while the RPP-stabilised slope remains stable and performs satisfactorily without any serviceability problems.

Because of its sustainable, resilient and cost-effective nature (cost savings compared to the conventional method vary between 50 and 80%), the use of RPPs for slope stabilisation has received considerable attention from many state departments of transportation (DOTs) and private companies. Moreover, the RPP-stabilised slopes in North Texas have attracted local, state and national media (both TV and newspaper) attention because of their sustainable nature. Recently, in March 2016, the editorial board of the Star-Telegram (one of the oldest newspapers in North Texas) stated, 'We need more solutions like this'.

It is understood that currently there is no book available on the market that provides design and construction guidelines for slope stabilisation using RPPs. This provided the motivation to work on this book to present design guidelines in the use of this green technology for slope stabilisation to professionals in both the USA and abroad. Taken in its entirety, the current book presents a robust and easy-to-use design method which will help both professional geotechnical engineers and the research community to design their slope stabilisation/repair solutions using RPPs. Case studies from both Texas and Missouri of successful slope stabilisations using RPPs are included in the book. Sample hand calculations for designing slope stabilisation using RPPs are also included in the book.

The authors would like to take the opportunity to thank everyone who helped make possible the implementation of slope stabilisation with RPPs in North Texas and who helped them complete this book. Special thanks to:

- Mr Bill Hale (who was District Engineer for TxDOT's Dallas District during the first implementation of an RPP-stabilised slope in Texas, 2011) for his unwavering support and confidence in UTA's research team, led by Dr Sahadat Hossain.
- Mr. Brian R. Barth, P.E., (District Engineer for the TxDOT's Fort Worth District) for his support and confidence in UTA's research team, led by Dr. Sahadat Hossain.
- Dr Nicasio Lozano, Mr Abbas Mehdibeigi, Mr Al Aramoon and Mr Boon Thian of TxDOT Dallas District for their help and support during the implementation of the projects.
- Mr Bill Pierce, area engineer of TxDOT Dallas District, for providing a site in his area to implement the first slope stabilisation project using RPPs.
- Those previous graduate students who were directly or indirectly involved during the implementation of the project, and who are still working on ongoing projects: Jubair Hossain, Shahed Rezwan Manzur, Mahsa Hedayati, Sandip Tarmakar, Mohammad Ashrafuzzaman, Asif Ahmed, Jobair Bin Alam, MD Faysal and Aayush Raj Tiwary. We are extremely thankful to them for their great service.
- Those current graduate students who helped us with different aspects of this book: Naima Rahman, Umme Zakira, Saif Bin Salah and Sangeeta Bhattacharjee. We are very thankful to them for their hard work, dedication and great service.

- Dr Sonia Samir, SWIS Manager, for her help and support during the completion of this book.
- Vance Kemler, General Manager, and David Dugger, Facility Manager, Solid Waste & Recycling Services at the City of Denton, for their continuous support for sustainable waste management research and their encouragement to complete this book.
- Ms Ginny Bowers, our official reviewer/editor, for checking grammar and other aspects of the book.
- Professor John Bowders of the University of Missouri, Columbia, for his gracious support and permission to include the case studies from Missouri in our book.
- Dr Janjaap Blom, editor at the Taylor and Francis Group, for his keen interest in the book.

Sahadat Hossain, PhD, PE
Sadik Khan, PhD, PE
Golam Kibria, PhD, PE

About the authors

Dr. Sahadat Hossain is a Professor of Civil Engineering Department and Director of Solid Waste Institute for Sustainability (SWIS) at the University of Texas at Arlington. Dr. Hossain has more than 20 (twenty) years of professional and research experience in geotechnical and geo-environmental engineering. Dr. Hossain's research experience includes slope stability analysis, innovative slope stabilisation techniques, assessment of geo-hazard potential, and recycled aggregate materials for base and sub-base applications, pavement crack mitigation, and sustainable waste management. One of his most recent successfully implemented projects is the slope stabilisation with recycled plastic pins, which is a major signature project in Texas. He also worked on more than 150 (One Hundred and Fifty) geotechnical design and construction projects in Bangladesh, Singapore, Hong Kong, Malaysia, Thailand and USA as a Civil/Geotechnical engineer. His working experiences include foundation analysis and design for building and bridge, excavation support system and retaining structures, cut and cover tunneling, slope stability analysis, design and construction of drilled shaft, contiguous bored pile wall, secant pile wall and diaphragm wall. Dr. Hossain received his bachelor's degree from Indian Institute of Technology (IIT), Bombay, India and Master's degree in Geotechnical Engineering from Asian Institute of Technology (AIT), Bangkok, Thailand. He received his PhD degree in Geo-Environmental Engineering from North Carolina State University (NCSU) at Raleigh, USA.

Dr. Sadik Khan is an Assistant Professor at the Department of Civil and Environmental Engineering at Jackson State University (JSU). Dr. Khan has completed his MS on August 2011 and PhD in Geotechnical Engineering on December 2013, from the University of Texas at Arlington (UTA). Before his graduate studies, he worked in several design and construction projects in Bangladesh. During his PhD dissertation, he conducted extensive research in the slope stabilisation using recycled plastic pin in Texas. He also worked in different research projects on failure investigation of highway slopes and MSE wall and geophysical evaluation of existing highway sub-structures. He has extensive experience in the instrumentation of highway slopes and pavement sites and also very competent in numerical studies using Finite Element Method (FEM). Dr. Khan is a Professional Engineer (P.E.) in the State of Texas.

Dr. Golam Kibria is a Geotechnical Engineer at Arias Geo-professionals, Inc., Texas. He received his Master's and Ph.D. degree from the University of Texas at Arlington. He has work experience in Bangladesh and United States. He has been working with different consulting companies in California and Texas. His research and work experience include: subsurface investigation using conventional and geophysical methods (i.e. electrical resistivity imaging, seismic survey), slope stability analysis, remediation of failed slope, foundation design, earth retention system design, and pavement design. He extensively worked with finite element modeling (FEM) and limit equilibrium analyses to evaluate slope stability, failure analysis of MSE walls, and design of laterally loaded drilled piers. Dr. Kibria is a Professional Engineer in the State of Texas.

Chapter 1

Introduction

Slope failures and landslides cause significant hazards in both public and private sectors. Each year, millions of dollars are spent on the repair and maintenance of highway slopes in the United States. Moreover, there are indirect costs and loss of revenue associated with landslides, which sometimes exceed the direct cost (Turner and Schuster, 1996). Costs associated with routine maintenance and repair of 'surficial' slope failures also represent significant, but generally neglected, losses due to landslides (Loehr and Bowders, 2007). Surficial slope failure is generally characterised by a failure depth of less than ten feet. Although the depth and area of surficial failure vary with soil type and slope geometry, failure depths of three to six feet are very common (Loehr and Bowders, 2007). Turner and Schuster (1996) conservatively estimated the cost of shallow slope failures as equal to, or even greater than, the costs associated with the repair of major landslides. In addition, shallow failures often cause significant hazards to infrastructures, such as guard rails, shoulders, and portions of roadways. Without proper maintenance, even more extensive and costly repairs are required (Loehr et al., 2007).

In general, slope failure occurs when the resisting force along the failure plane is less than the driving force. The possibility of slope failure increases when cut or fill slope is underlaid by expansive clay. More than 25% of the total area of the United States is underlaid by highly plastic clay soil, which has moderate-to-high swelling potential, as presented in Figure 1.1 (Nelson and Miller, 1992). Moderate-to-steep slopes and embankments underlaid by expansive clayey soils are known to be susceptible to shallow landslides during intense and prolonged rainfall events. During or after rainfall, failure occurs due to an increase in pore water pressure and reduction in soil strength at the shallow subsurface. This condition is further exacerbated by moisture variations due to seasonal climatic changes, and results in cyclic shrinkage and swelling at shallow depths. The shrinkage cracks act as a conduit for surface water infiltration from rainfall (McCormick and Short, 2006; Hossain et al., 2016; Khan et al., 2016). Due to the wetting-drying cycle, sloughing and shallow slope failures are predominant in many parts of the United States and pose a significant maintenance problem, as depicted in Figure 1.2.

The conventional slope stabilisation technique includes the installation of a drilled shaft, replacement of the slope using a retaining wall, installation of soil nails, and reinforcement of the slope using geogrids. These techniques for the repair of shallow slope failure can be expensive and time-consuming; therefore, new, innovative, sustainable

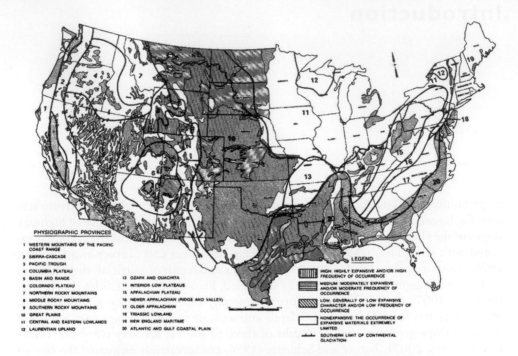

Figure 1.1 Expansive clay map of the United States of America (Department of the Army, 1983).

Figure 1.2 Typical slope failure in the North Texas area.

and cost-effective slope stabilisation methods are being tried, tested, and increasingly implemented. One such method is the use of RPPs.

Globally, 1.3 billion tons of MSW are generated annually, and this annual amount is expected to increase to 2.2 billion tons by 2025. The average plastic waste is around

10% of the total waste, which amounts to approximately 130 million tons. In the United States, 13% of generated MSW is plastics, which is approximately 32.5 million tons. However, the rate of plastic recycling is 30% in Europe, but only 10% in the USA (PlasticsEurope, 2013). An additional 30% of plastics waste is used for waste-to-energy plants in Europe. The rest of the plastics (90% in USA and 40% in Europe) are discarded into landfills.

Plastic and plastic products are a non-degradable component of the MSW stream. Even though they are lightweight, they occupy a large volume of landfill space. As they are non-degradable, they remain in the landfill for long periods and occupy valuable landfill space which otherwise could have been reused for additional waste burial. Therefore, diverting and/or reusing plastic products from landfills could save landfill space in the short term and help stabilise the landfill at a faster rate in the long term.

The use of RPPs, made out of recycled plastic bottles, for highway slope stabilisation and civil engineering infrastructure projects demonstrates a perfect example of sustainable engineering solutions. Plastic, being non-degradable, is problematic if discarded into landfills; however, due to its non-degradable nature, it could be very useful and beneficial for civil engineering infrastructure projects. Soil slopes and highway soil slopes repaired with RPPs preserve their engineering characteristics for a long time, thereby reducing the overall maintenance and repair costs of slopes and pavement shoulders over time.

RPPs have been utilised in Missouri and Iowa as a cost-effective solution for the stabilisation of shallow slope failure (Loehr and Bowders, 2007). They are predominantly a polymeric material, fabricated from recycled plastics and other waste materials (Chen et al., 2007; Bowders et al., 2003). RPPs are composed of high-density polyethylene, HDPE (55–70%), low-density polyethylene, LDPE (5–10%), polystyrene, PS (2–10%), polypropylene, PP (2–7%), polyethylene terephthalate, PET (1–5%), and varying amounts of additives, e.g. sawdust, fly ash (0–5%) (Chen et al., 2007; McLaren, 1995; Lampo and Nosker, 1997).

RPPs are manufactured primarily by using recycled plastic bottles; approximately 600 soda bottles are used for one RPP (Figure 1.3). It is a lightweight material and is less susceptible to chemical and biological degradation, more resistant to moisture, and requires very little maintenance, thus making it a more attractive alternative than other structural materials (Krishnaswamy and Francini, 2000).

RPPs are installed in the slope to intercept potential sliding surfaces, which provides additional resistance to maintain the long-term stability of the slope. A number of studies have been conducted over the past few years in an effort to stabilise shallow slope failures by using RPPs (Khan et al., 2016; Khan et al., 2013; Loehr and Bowders, 2007). The study results indicate that this slope stabilisation method has great potential as an effective and green technique. A limit state design technique is available in the literature that considers the resistance from the RPPs in inhibiting shallow slope failure (Loehr and Bowders, 2007). However, the current design methods that are used by geotechnical engineers have some limitations:

1   The design does not consider deformation of RPPs and soil slopes during the design life of the RPP-stabilised slopes.
2   The creep behaviour of RPPs is not considered in the design.

Figure 1.3 Recycled plastic pins.

Because of the cost-effective and sustainable aspects of slope stabilisation using RPPs, this stabilisation method is getting enormous attention from both the public and private sectors. More and more state DOTs and private companies are now considering the use of RPPs for slope stabilisation. Hence, it becomes important to develop a design method that takes account of current limitations.

To the knowledge of the authors, there is currently no book available that provides design and construction guidelines for slope stabilisation using RPPs. Design protocols and design charts that connect different important parameters, such as the length and spacing of RPPs, the height and angle of slopes, and the soil parameters, are imperative in bridging the gap between researchers, geotechnical engineers and contractors. This book presents a comprehensive and easy-to-use design method which will help both the research community and professional geotechnical engineers to design their slope stabilisation/repair solutions using RPPs.

The design charts presented in this book were developed using the finite element (FE) program, PLAXIS (2011). First, the FE program was calibrated, using field monitoring results and field performance data, after which the design charts were developed using different design parameters. Although the limit equilibrium method is widely used in slope stability analyses, the FE method has the ability to provide a serviceability-based design. The graphical presentation of the FE program allows better understanding of the failure mechanism (Griffiths and Lane, 1999). The shear strength reduction criteria of the FE method have been successfully used to evaluate the stability of slopes reinforced with piles and anchors (Wei and Cheng, 2009; Yang et al., 2011; Cai and Ugai, 2003). This technique was also used to evaluate the stability of slopes reinforced with piles or anchors under a general frame, where soil–structure interactions were considered, using zero-thickness elasto-plastic interface elements (Chen and Poulos, 1993; Chen and Poulos, 1997).

At present, no study is available which demonstrates slope stabilisation using RPPs together with the serviceability aspects of this method. In this book, finite element modelling (FEM) is effectively utilised to observe the influence of different lengths and spacings of RPPs to determine the factor of safety and deformation of the stabilised slope.

The book is organised as follows:

Chapter 1 provides a background to slope failure and slope stabilisation techniques.

Chapter 2 presents the mechanism of shallow slope failure and various techniques to repair the failed slope.

Chapter 3 describes the generation and recycling of plastic waste and the corresponding environmental impacts.

Chapter 4 includes a review of the manufacturing method, physical properties, strength and modulus of RPPs for structural applications and for slope stabilisation. The chapter presents an experimental study to determine the physical properties of RPPs with different loading rates.

Chapter 5 describes the development of a design method for slope stabilisation using RPPs. The design method considers three limiting criteria, which include the failure of the soil adjacent to the RPP, as well as the horizontal displacement and creep stress of the RPP. The chapter introduces the design steps and compares the FE method with the limit equilibrium method.

Chapter 6 presents the details of the construction technique, the equipment suitable for construction, potential challenges, and special installation techniques.

Chapter 7 summarises five case studies of slope stabilisation using RPPs. Each of the case studies includes descriptions of the site investigation, the design of the slope stabilisation technique and the design layout, installation, and the results of performance monitoring.

The book is organized as follows.

Chapter 1 provides a background to slope failure and slope stabilization techniques.

Chapter 2 presents the mechanism of shallow slope failure and various techniques to repair the failed slope.

Chapter 3 describes the generation and recycling of plastic waste and the corresponding environmental impacts.

Chapter 4 includes a review of the manufacturing method, physical properties, strength and modulus of RPF, for structural applications and for slope stabilization. The chapter presents an experimental study to determine the physical properties of RPF, with different loading rates.

Chapter 5 describes the development of a design method for slope stabilization using RPF. The design method considers three limiting criteria, which include the failure of the soil adjacent to the RPF, as well as the horizontal displacement and creep strain of the RPF. The chapter introduces the design steps and compares the LE method with the limit equilibrium method.

Chapter 6 presents the details of the construction technique, the equipment, suitable for construction, potential challenges, and special installation techniques.

Chapter 7 summarizes five case studies of slope stabilization using RPF. Each of the case studies includes descriptions of the site investigation, the design of the slope stabilization technique and the design layout, installation, and the results of performance monitoring.

Chapter 2

Slope failure and stabilisation methods

2.1 SLOPE FAILURE

Soil slope failures are common occurrences globally and usually occur after prolonged rainfall events, which lead to the reduction of soil strength (Titi and Helwany, 2007). Slope failures can take place gradually, or can happen without any warning. Slopes are generally characterised as stable when the shear strength of the soil provides adequate resistance against the gravitational forces that are trying to move the soil mass down the slope. Therefore, the stability of the slope is governed by the balance between the driving and resisting forces. Changes in these forces may lead to the loss of stability and subsequent slope failure. An increase in driving (gravitational) forces can be triggered by changes in slope geometry, seepage pressure, or added surcharge from traffic loads on highway embankments (Titi and Helwany, 2007). On the other hand, reduction in resisting forces can take place due to increased pore water pressure (as water perches on impermeable underlying soil layers) and decreased soil shear strength, due to the saturation of clayey soil after a prolonged rainfall event.

Slopes and embankments constructed on highly plastic clay soil are usually very strong immediately following construction. Skempton (1977) first reported that, over time, the strength of slopes in the highly plastic London clay decreases, eventually reaching what Skempton termed a 'fully softened' strength. A few years after construction, shrink–swell behaviour can reduce the shear strength of the top few feet of a slope, which is susceptible to moisture variations. After the soil reaches the fully softened shear strength, water that infiltrates the soil during intense and prolonged rainfall events may cause excess pore water pressure and compound the overall problem. As a result, the factor of safety of the slope may reduce to unity and approach failure (Wright, 2005). These types of failures occur within the upper one to two metres (3–6 ft) of the slope, and the failure surface is generally parallel to the slope face (Wright, 2005). Surficial failure may occur within three to seven years of construction, but can take decades to form depending on the frequency of extreme weather conditions.

Slopes fail when the soil mass between the slope and slip surfaces move towards the downslope. According to Titi and Helwany (2007), the soil movement and depth of the slip surface depends on the type of soil, soil stratification, slope geometry and the presence of water. Abramson et al. (2002) described typical slides that can occur in clay soils, such as (1) translational, (2) plane or wedge surface, (3) circular, (4) noncircular, and (5) a combination of these types. The different slope failure types are illustrated in Figure 2.1.

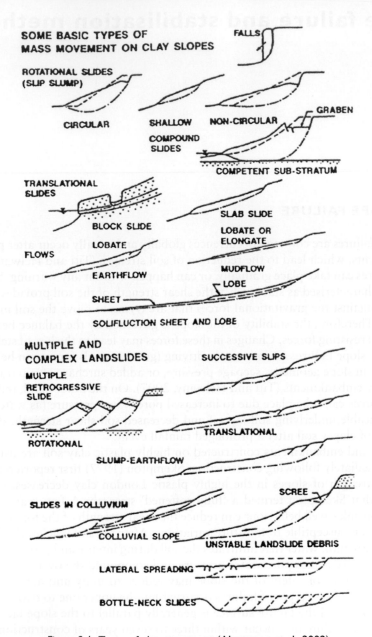

Figure 2.1 Types of clay movement (Abramson et al., 2002).

Design of a stable slope requires a rational selection and use of a factor of safety that accounts for the various uncertainties associated with the determination of soil strength, distribution of pore pressures and soil stratification. It is suggested that a high factor of safety of slope be considered when the level of soil investigation is of low quality and the experience of the engineer is limited (Abramson et al., 2002).

The factor of safety of a slope is calculated by comparing the available shear strength along a potential slipping plane with the equilibrium shear stress that is needed to maintain a barely stable slope. The factor of safety is assumed to be constant along the slip surface and can be defined in terms of stresses (total and effective), forces and moments. Selecting a factor of safety for a typical slope design depends on many factors, including the level and accuracy of soil data, the experience of the design engineer and the contractor, the level of construction monitoring, and the consequence of slope failure (risk level) (Titi and Helwany, 2007). For a typical slope design, the required factor of safety ranges between 1.25 and 1.50 (Abramson et al., 2002).

2.2 SHALLOW SLOPE FAILURE

Surficial failures of slopes are quite common throughout the United States. Shallow slope failure refers to surficial slope instabilities along highway cuts, fill slopes and embankments. These instabilities commonly occur in fine-grained soils, especially after prolonged rainfall. Surficial failure, by definition, is shallow, with the failure surface usually at a depth of 1.2 m (4 ft) or less (Day and Axten, 1989). In many cases, the failure surface is parallel to the slope face, as illustrated in Figure 2.2.

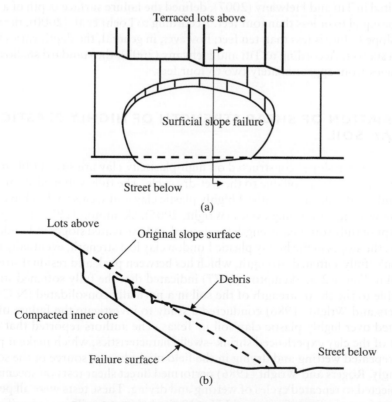

Figure 2.2 Typical surficial slope failures (Day, 1989).

Shallow slope failures generally do not constitute a hazard to human life or cause major damage. However, they can constitute a hazard to infrastructure by causing damage to guardrails, shoulders, road surfaces, drainage facilities, utility poles, or the slope landscaping (Khan et al., 2015; Khan et al., 2016; Hossain et al., 2016). In some cases, shallow slope failures can impact traffic flow if debris flows onto highway pavements. Moreover, shallow slope failures can have an economic impact on highway agencies at a local/district level. In general, the repairs of shallow slope failures are conducted at the local district level and are often performed by maintenance crews as routine maintenance work. In many cases, such repairs may provide only a temporary fix, as the slope failure generally reoccurs after a rainfall season (Titi and Helwany, 2007).

In general, shallow slope failures vary in depth and extent of the failed area and these variances largely depend on slope geometry, soil type, degree of saturation of soil, seepage, and climatic conditions (Titi and Helwany, 2007). According to Abramson et al. (2002), many shallow slope failures occur when the rainfall intensity is greater than the soil infiltration rate and the rainfall lasts long enough to saturate the slope up to a certain depth, which leads to the build-up of pore water pressure at that depth.

Shallow slope failures are often parallel to the slope surface and are usually considered infinite slope failures. Various depths were reported in the literature, based on case histories, but all studies indicated the shallow nature of surficial failures. Evans (1972), cited in Titi and Helwany (2007), defined the failure surface depth of a shallow slope to be equal to or less than four feet. According to Loehr et al. (2000), the depth of shallow slope failure is less than ten feet; however, in general, the depth varies between three and six feet. According to Titi and Helwany (2007), the standard shallow failure depth ranges from approximately two to four feet.

2.3 VARIATION OF SHEAR STRENGTH OF HIGHLY PLASTIC CLAY SOIL

Moderate-to-steep slopes constructed on high-plasticity clay are susceptible to softening behaviour at the top soil due to the wet-dry cycle. The fully softened shear strength corresponds to the shear strength of highly plastic clay and seems to develop over time due to the wetting and drying cycles (Wright, 2005). Skempton (1977) first proposed the concept of fully softened strength for natural and excavated slopes in London clays: over time, the slopes in the highly plastic London clay lost strength, eventually reaching Skempton's 'fully softened' strength, which lies between peak and residual strength, as presented in Figure 2.3. Skempton (1977) indicated that the fully softened strength is comparable to the shear strength of the soil in a normally consolidated (N-C) state.

Rogers and Wright (1986) conducted a study to investigate the failure of a slope constructed over highly plastic clay soil in Texas. The authors reported that the high plasticity of the clay experienced shrink–swell characteristics, which make it probable that the repeated wetting and drying in the field might be one source of the softening. Accordingly, Rogers and Wright (1986) performed direct shear tests on specimens that were subjected to repeated cycles of wetting and drying. These tests were all performed on the clay from the Scott Street and I.H. 610 site in Houston, Texas, identified as red

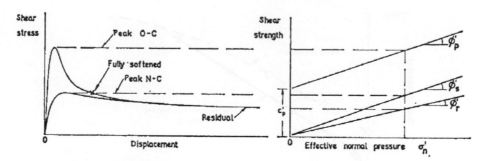

Peak O-C: Peak Over Consolidated
Peak N-C: Normally Consolidated

Figure 2.3 Comparisons of peak, residual and fully softened shear strength (Skempton, 1970).

Table 2.1 Summary of shear strength parameters from drained direct shear tests on specimens subjected to wetting and drying cycles (Rogers and Wright, 1986).

Number of wet-dry cycles	Cohesion, c (psf)	Friction angle, φ
1	29	23°
3	77	26°
9	33	25°
30	0	27°

clay. Four series of drained direct shear tests were performed on specimens subjected to 1, 3, 9 and 30 cycles of wetting and drying. The shear strength parameters obtained from the study are summarised in Table 2.1.

Rogers and Wright (1986) reported that cyclical wetting and drying of the soil produces a significant shear strength loss, particularly in terms of effective cohesion intercept, c'. The direct shear test also indicated that the loss in cohesion occurs within a relatively few cycles of wetting and drying. In fact, most of the loss in strength occurred on the first cycle. However, Rogers and Wright suggested that the wetting and drying to which specimens were subjected in the laboratory was much more severe than those expected to occur in the field. Even so, the effects of wetting and drying in the laboratory and field are believed to be similar.

Kayyal and Wright (1991) developed a new procedure for triaxial specimens subjected to repeated cycles of wetting and drying. The procedure allowed the specimens greater access to moisture and exposure for drying. In addition, the procedure allowed substantial lateral expansion and volume change to occur in the soil during drying. Two soils were tested during the study: red or Beaumont clay from Houston, Texas, and highly plastic clay soil from Paris, Texas.

Kayyal and Wright (1991) conducted several series of consolidated undrained compression tests with pore pressure measurements. Tests were performed on specimens

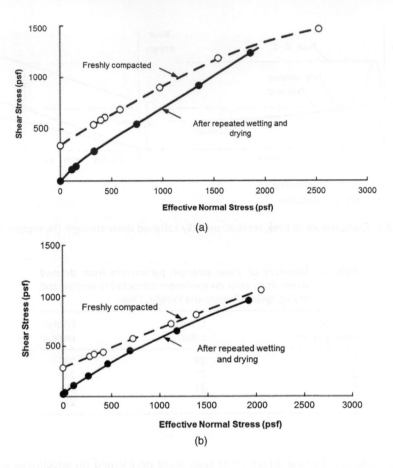

Figure 2.4 Shear strength envelopes in terms of effective stress: a. Beaumont clay; b. Paris clay (Kayyal and Wright, 1991).

subjected to repeated wetting and drying, as well as on freshly compacted samples. Based on the study, the shear strength envelopes for specimens of Beaumont clay and Paris clay were tested as compacted after wetting and drying, and the results are presented in Figure 2.4. The results show that both envelopes were distinctly non-linear. In addition, at lower values of normal stress, the strength envelope of the specimens subjected to wetting and drying cycles fell significantly below the envelope of the specimens tested in the compacted condition. Moreover, the intercept of the strength envelope for specimens subjected to wetting and drying was small and could be considered negligible.

Slopes constructed in arid and semi-arid regions remain in an unsaturated condition above the groundwater table. To accurately predict the mechanical behaviour of unsaturated soil, two stress state variables are required, the most widely used of which is the combination of net normal stress $(\sigma - u_a)$ and matric suction $(u_a - u_w)$. Based on these two stress state variables, Fredlund et al. (1978) proposed the following

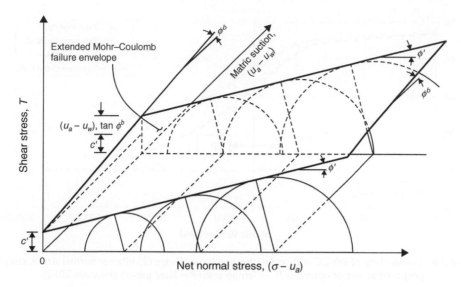

Figure 2.5 Extended Mohr–Coulomb failure envelope for unsaturated soils (Fredlund et al., 2012).

equation, which is an extension of the Mohr–Coulomb (MC) theory, to describe the shear strength of unsaturated soil:

$$\tau = c' + (\sigma - u_a)\tan\varphi' + (u_a - u_w)\tan\varphi^b \tag{2.1}$$

where $(\sigma - u_a) =$ net normal stress, $(u_a - u_w) =$ matric suction, $\varphi' =$ angle of internal friction associated with the change in net normal stress, and $\varphi^b =$ angle representing the rate of change in shear strength relative to matric suction change.

The first two terms on the right of Equation 2.1 describe the conventional MC theory to determine the strength of saturated soil. The third term indicates the change in shear strength due to the change in matric suction in the unsaturated soil. The corresponding failure envelope for the extended MC criterion is presented in three-dimensional stress space in Figure 2.5.

Hossain (2012) conducted a study at the University of Texas at Arlington to investigate the behaviour of rainfall-induced slope failure over expansive soil in Texas. During the study, the unsaturated soil behaviour of the highly plastic clay soil was determined, using both laboratory tests and field investigations.

Highly plastic clay soil is subjected to swelling and shrinkage behaviour when exposed to moisture variations. This volume change also affects the void ratio of the soil, which impacts the matric suction. Hossain (2012) considered the volume change behaviour of the expansive soil and determined the soil-water characteristic curve (SWCC) of highly plastic compacted clay soil specimens, considering both volume change and no-volume-change (NVC) behaviour. The SWCCs of the specimens compacted wet at an optimum of 25 kPa net normal stress, in terms of both volume change and NVC, are presented in Figure 2.6. In both cases, the experimental results were fitted with the van Genuchten (1980) model. Hossain (2012) determined that the SWCC that considered volume change showed a higher air entry value (100 kPa)

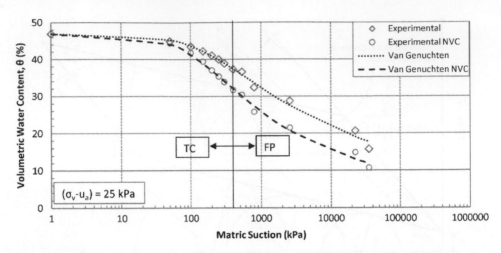

Figure 2.6 Comparison of SWCC with and without volume change (25 kPa net normal stress, samples prepared at wet of optimum; *TC* = Tempe cell; *FP* = filter paper) (Hossain, 2012).

Table 2.2 Best-fitting van Genuchten parameters with and without volume change at wet of optimum (Hossain, 2012).

Parameters	25 kPa		50 kPa		100 kPa	
	VC	NVC	VC	NVC	VC	NVC
α (1/kPa)	0.0064	0.0089	0.0063	0.0096	0.0064	0.0072
n	1.219	1.3116	1.2184	1.3009	1.2354	1.3122
m	0.1797	0.2376	0.1792	0.2313	0.1908	0.2379

Table 2.3 Best-fitting van Genuchten parameters with and without volume change at dry of optimum (Hossain, 2012).

Parameters	25 kPa		50 kPa	
	VC	NVC	VC	NVC
α (1/kPa)	0.0159	0.0123	0.0123	0.0134
n	1.1454	1.2407	1.1335	2.335
m	0.1269	0.1940	0.1178	0.0833

than the NVC one (60 kPa). Moreover, the slope of the curve is steeper in the latter case, compared to the volume change curve, which over-predicts the desaturation rate of the soil. The van Genuchten fitting parameters, taking into account volume change and NVC for samples prepared on the wet side (wet of optimum) and dry side (dry of optimum) of optimum moisture content, are presented in Tables 2.2 and 2.3, respectively.

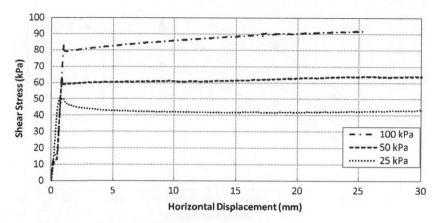

Figure 2.7 Multistage ring shear test under matric suction $(u_a - u_w) = 50$ kPa and net normal stresses of 25, 50 and 100 kPa (Hossain, 2012).

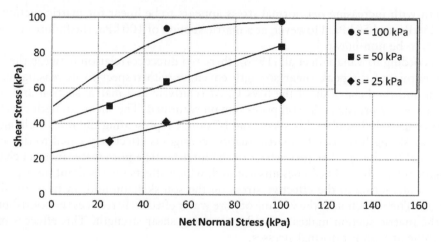

Figure 2.8 Effect of matric suction on the peak failure envelope of expansive clay (Hossain, 2012).

Hossain (2012) conducted a multistage shear test, using a suction-controlled ring shear device in order to study the influence of suction on the shear strength of unsaturated Eagle Ford clay. Three tests were carried out at suction states $(u_a - u_w) = 25$, 50 and 100 kPa, under three different net normal stresses $(\sigma - u_a) = 25$, 50 and 100 kPa. The effect of net normal stresses on the shear strength for a suction of 50 kPa is shown in Figure 2.7. A clear peak was observed, which subsequently reached a residual state condition, indicating a hardening-softening behaviour. Moreover, both the peak and residual strength increased with the increase in net normal stresses.

Figure 2.8 presents the effects of matric suction on both the position and slope of the unsaturated *peak* failure envelope of compacted high plasticity index (PI) expansive clay, based on the multistage ring shear test results. It can be observed that matric suction plays a significant influence on the position of the peak failure envelope, with a considerably higher position for a suction value of 100 kPa. The increase in peak

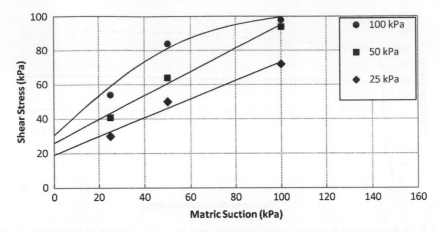

Figure 2.9 Effect of matric suction on the residual failure envelope of expansive clay (Hossain, 2012).

strength with increasing net normal stress appears to be linear for matric suctions of both 25 kPa and 50 kPa. However, at a matric suction of 100 kPa, the failure envelope appears to be non-linear.

According to Vanapalli et al. (1996), there is a direct correlation between the non-linear nature of the peak shear strength envelope with respect to increasing matric suction and the behaviour of the SWCC. At relatively low matric suction and prior to the air entry pressure, the soil pores remain saturated. The shear strength envelope is approximately linear at this stage, and φ^b (angle representing the rate of change in the shear strength relative to matric suction change) is effectively equal to the angle of internal friction φ'. The geometries of the inter-particle pore water menisci change dramatically as the soil becomes unsaturated, which affects the resultant inter-particle forces contributing to the effective stress on the soil skeleton and, in turn, on shear strength. The reduction in the volume of pore water effectively reduces the contribution that the matric suction makes towards increasing shear strength. This effect is more noticeable at high net normal stresses.

The effect of matric suction on both the position and slope of the unsaturated *residual* failure envelopes of compacted high PI expansive clay, according to the multistage ring shear test conducted by Hossain (2012), is presented in Figure 2.9. It can be observed that matric suction has a significant influence on the position of the residual failure envelope, with a considerably higher position for the suction value of 100 kPa. For the matric suction range considered, the increase in residual strength, with increasing net normal stress, appears to be linear. However, at 100 kPa suction, the failure envelope appears to be non-linear. It should be noted that the change in shear strength for residual soil follows a trend similar to that of the peak strength (presented in Figure 2.8).

Hossain (2012) also conducted multistage direct shear tests on saturated specimens to study the effect of softening behaviour due to the wetting and drying of Eagle Ford shale. During this study, he conducted direct shear tests using the conventional approach to sample preparation (Test 1), sample preparation of expansive soil involving NVC (Test 2), and sample preparation of expansive soil allowing volume change

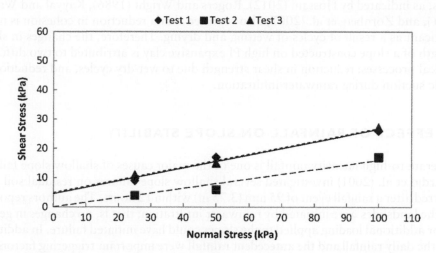

Figure 2.10 Effect of wetting and drying on peak failure envelope of expansive clay (Hossain, 2012).

Table 2.4 Peak shear strength parameters for multistage direct shear tests (Hossain, 2012).

Test no.	Normal stress (kPa)	Wet-dry cycles	No. of stages	Condition	c′ (kPa)	φ′
I	25 50 100	0	3	Conventional	4.515	12.41
2	25 50 100	I	3	NVC	0	9.09
3	25 50 100	I	3	VC	4.92	12.37

(Test 3). The failure envelopes for the three tests performed under different conditions are presented in Figure 2.10. From the figure, it can be observed that Tests 1 and 3 have almost identical failure envelopes. On the other hand, reduction in the shear strength parameter was observed for Test 2. Table 2.4 presents the estimated shear strength parameters for the three multistage tests performed under the different conditions. Based on Table 2.4, it can be seen that Test 2, which was performed with one cycle of wetting and drying, allowing no volume change, showed a decrease in both cohesion intercept and angle of internal friction. On the other hand, no change in shear strength was observed in Test 3, which was also subjected to one cycle of wetting and drying but allowed volume change. This is in contrast to Test 1, which was performed without the soil being subjected to a wet/dry cycle.

Based on the test results, it can be observed that the reduction in cohesion due to cycles of wetting and drying is more significant than the reduction in friction angle. The shear strength of soil derived from cohesion was completely lost after wet-dry cycles, as indicated by Hossain (2012). Rogers and Wright (1986), Kayyal and Wright (1991), and Zornberg et al. (2007) also reported that a reduction in cohesion is more significant as a result of cycles of wetting and drying. Therefore, the changes in shear strength of a slope constructed on high PI expansive clay is attributed to two different physical processes: reduction in shear strength due to wet-dry cycles, and reduction in matric suction during rainwater infiltration.

2.4 EFFECT OF RAINFALL ON SLOPE STABILITY

Moderate-to-high intensity rainfall is one of the major causes of shallow slope failure. Rahardjo et al. (2001) investigated several shallow slope failures on residual soil that occurred after a rainfall event of 95 mm (3.75 in) within 12 hours. The authors reported that the landslides were initiated by rainwater infiltration; that is, no changes in geometry or additional loading applied to the slopes could have initiated failure. In addition, both the daily rainfall and the antecedent rainfall were important triggering factors for the occurrence of the landslides. Rahimi et al. (2010) conducted a study on rainfall-induced slope failure due to antecedent rainfall for high- and low-conductivity residual soils in Singapore. The authors applied three antecedent rainfall patterns to soil slopes and conducted a transient seepage analysis to investigate the effect of rainfall on the stability of the slope. Results from the study indicated that antecedent rainfall affected the stability of both high- and low-conductivity soil slopes, with the stability of the low-conductivity soil slopes being more significantly affected. In addition, the stability of the slope was controlled by the amount of rainfall that infiltrated the unsaturated zone of the slope.

Hossain (2012) instrumented a highway slope in Dallas, Texas, constructed on highly plastic clay soil. During the study, moisture sensor and water potential probes were installed at the crest of the slope to measure the moisture variations and associated changes in matric suction due to rainfall. The study measured the moisture and suction variations between August 2010 and May 2012. Based on the study, variations of *in situ* moisture content and matric suction profiles at 1.2 m (3.6 ft) depth are presented in Figures 2.11a and 2.11b, respectively. The initial moisture content was relatively low (4%), and matric suction was high (−600 kPa). However, a total precipitation of 181.9 mm (7.2 in) was observed in a period of one month between April and May 2011, with the highest daily precipitation of 54.2 mm (2.1 in) during two rainfall events in April and May. As a result, the moisture content increased to a maximum of 37%, and the matric suction decreased to −10 kPa. Based on the back calculations of the failures of several slopes constructed on Paris and Beaumont clays, Aubeny and Lytton (2004) suggested that the wet limit of suction at the surface during a slope failure is 2 pF (i.e. approximately −10 kPa).

During the relatively dry summer of 2011 in Texas, the moisture content decreased gradually from 37 to 10%; accordingly, matric suction increased from −100 to −450 kPa. Therefore, it is clear that at a depth of 1.2 m (3.6 ft), the variation of moisture content due to rainfall is evident. Moisture content in the soil began to increase with the rainfall observed between September and December 2011, but the matric suction in the soil remained unchanged. This may be due to changes in the soil moisture retention characteristics of the upper part, as a result of cracking during the dry period.

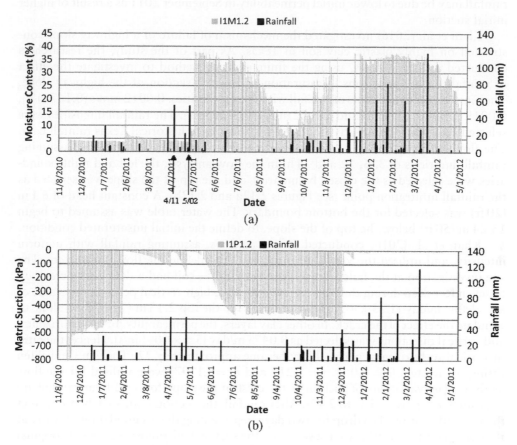

Figure 2.11 *In situ* variation of moisture content (a) and matric suction (b) at 1.2 m (3.6 ft) depth near the crest of a highway slope in Dallas, Texas (Hossain, 2012).

Zornberg et al. (2007) also reported significant changes in soil moisture retention characteristics due to the formation of cracks in high-plasticity clays. The total rainfall observed during this period was 176 mm (6.9 in), with the highest daily observed rainfall being 29 mm (1.1 in). Moisture content at the crest of the slope at 1.2 m (3.6 ft) depth remained close to 37% between January and March 2012, as there was a high volume of rainfall during this period. The matric suction also remained unchanged, at a value of −10 kPa, due to the large volume of rainwater infiltration during this period. Hossain (2012) also observed that infiltration of rainwater depends on the initial matric suction of the soil. During a rainfall event in September 2011, a total of 26.4 mm (1.03 in) rainfall was observed, but both moisture content and matric suction remained unchanged. The *in situ* matric suction prior to the rainfall was approximately −450 kPa. On the other hand, the same amount of rainfall was registered in an event in December 2011, and a drop in suction was observed from −50 kPa (−7.2 psi) to approximately −10 kPa (−1.4 psi). The different behaviours for the same amount of rainfall may be due to lower initial permeability in September 2011 as a result of higher initial suction.

Khan et al. (2016) investigated the mechanism of failure of a highway slope constructed on highly plastic clay soil in Texas. As part of the study, the researchers conducted a flow analysis, using the finite element method to investigate the effect of rainfall on the variations in saturation and matric suction of the highway slope. They investigated the effects of rainfall with three intensities: 0.0012 mm/s (0.167 in/h), 0.0019 mm/s (0.271 in/h) and 0.0022 mm/s (0.3125 in/h). The rainfall intensities were selected according to two-, five- and ten-year periods of historical Texas rainfall data. The flow through the top soil was determined for each of the intensities, assuming rainfall durations of 3, 6, 12 and 24 h. In the flow analysis, the left and right boundaries were selected as the closed boundaries, and the top of the slope was selected as the rainfall infiltration point (see Figures 2.12a and 2.12b). A constant head of 6.1 m (20 ft) was selected for the bottom boundary. The water table was assumed to begin 15.24 m (50 ft) below the top of the slope, to define the initial unsaturated condition.

Khan et al. (2016) conducted a flow analysis, assuming rainfall with uniform intensity, and utilised the van Genuchten model to define the flow parameters. The site investigation at the failed slope indicated that the soil had a desiccation crack up to seven feet deep during the summer. As a result, a high vertical permeability value of $k_y = 1.063$ m/day (1.23×10^{-5} m/s) was used for the top 2.13 m (7 ft) of the slope to simulate the effect of the crack. In other clay layers, the permeability for both horizontal and vertical directions was selected as 0.0475 m/day (5.5×10^{-7} m/s). Hence, the ratio of vertical-to-horizontal permeability became approximately 22. The van Genuchten fitting parameters ($\alpha = 0.064$, $n = 1.219$ and $m = 0.1797$) were utilised for the flow analysis. The variations of suction at the crest of the slope for a 12-hour rainfall are presented in Figures 2.12c to 2.12f. After rainfall, the suction immediately dropped at the top and continued to drop for two days, representing the accumulation of water at the corresponding depth. After seven days, the suction had almost regained its original profile.

Based on the study conducted by Khan et al. (2016), the variations of change in suction at 0.91 m (3 ft), 1.82 m (6 ft) and 2.43 m (8 ft) depths are presented in Figure 2.13. The change of suction refers to the change from the initial suction value prior to rainfall. The figure shows various rainfall durations with a two-year return period (rainfall intensity of 0.0012 mm/s [0.167 in/h]) and a ten-year return period (0.0022 mm/s [0.3125 in/h]). The suction was observed to drop significantly with rainfalls of higher intensity and longer duration. Moreover, the change in suction was more significant at depths of 0.91 m (3 ft) and 1.82 m (6 ft) than at the greater depth of 2.43 m (8 ft).

The drop in suction was instantaneous at the depth of 1.82 m (6 ft) for different rainfall intensities and durations, as depicted in Figures 2.13b and 2.13e. In contrast, post-rainfall, the change in suction took several days to several weeks to achieve a steady value at 2.43 m (8 ft) depth. The constant value of suction at the 2.43 m (8 ft) depth indicated that the percolated water could not drain out of the slope due to the very low permeability of the highly plastic clay soil.

The flow pattern through the top soil was important to the investigation of the failure. The slope failed in October 2011, after a long drought condition followed by a rainfall event (Figure 2.13). Immediately after the rainfall, the existence of shrinkage cracks allowed for easy rainwater intrusion at the slope crest in the top few metres.

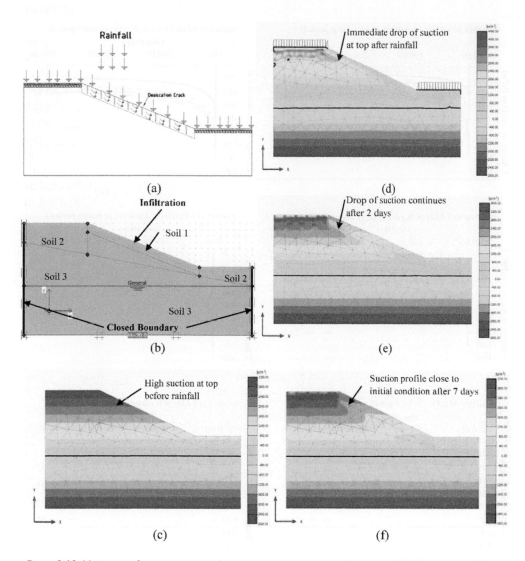

Figure 2.12 Variation of matric suction during moisture intrusion due to rainfall (Khan et al., 2016).

The low permeability of the highly plastic clay probably prevented downward movement of the water, creating a perched water zone near the crest of the slope. Hence, the slope stability analysis of the slope was extended to consider this zone.

Khan et al. (2016) integrated the flow analysis results of a failure investigation of a highway slope in Texas that was constructed over highly plastic clay soil. It was evident that, due to rainfall, water infiltrated and saturated the top soil to create a perched water condition near the crest. Therefore, a perched water zone was applied in the crest (top third of the slope face) to evaluate the effect of rainfall on the safety of the slope in the presence of desiccation cracks. During this study, Khan et al. (2016) also considered the fully softened shear strength of the top soil. The soil model with

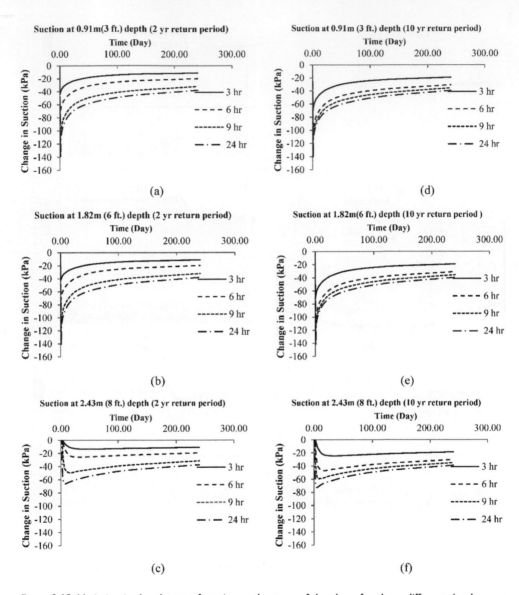

Figure 2.13 Variation in the change of suction at the crest of the slope for three different depths over two different return periods (Khan et al., 2016).

the perched water zone is presented in Figure 2.14. A phi-c reduction analysis was conducted, and the results indicated that the factor of safety of the slope reduced to 1.05, which is very close to failure, when the perched water condition was taken into consideration. The failure plane of the slope is presented in Figure 2.14. Khan et al. (2016) concluded that the slope failure took place due to the combined actions of the perched water zone at the crest, formed by rainwater intrusion through desiccation

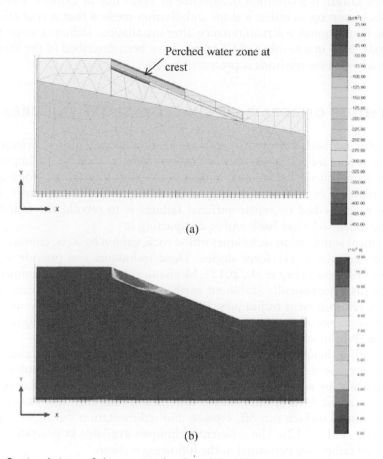

Perched water zone at crest

(a)

(b)

Figure 2.14 Sectional views of slope: a. initial soil model, including the perched water zone; b. slope stability analysis with fully softened strength and perched water zone at crest (Khan et al., 2016).

cracks following a prolonged summer, and the fully softened condition resulting from seasonal wet-dry cycles. The combination of the desiccation cracks and the perched water zone in the top few feet initiated the failure.

Wright (2005) reviewed various research projects on slopes and embankments constructed on highly plastic clay soil in Texas and reported that compacted, highly plastic fills are generally very strong immediately after construction. In general, at the end of construction, the factor of safety probably exceeds two, although the soils tend to soften and weaken over time. As a result, the factor of safety decreases to values that approach one, that is, failure. The softening is probably exacerbated by the repeated expansion and shrinkage which takes place due to seasonal wetting and drying of the soil, respectively. The fully softened strength of the soil is best characterised by a curved MC failure envelope. In addition, the failure envelope for fully softened conditions lies below the failure envelope for the soil immediately after compaction.

As slope failure is a common occurrence in Texas due to extreme weather conditions, it is important to utilise a slope stabilisation method that is cost-effective, as well as one that requires less maintenance after installation. Different slope stabilisation methods, used in a number of US states, have been described in the literature. A brief summary of these methods is presented below.

2.5 METHODS OF REPAIR OF SHALLOW SLOPE FAILURES

Various repair methods are used to stabilise surficial slope failures. Selection of an appropriate repair technique depends on several factors, such as the importance of the project (consequence of failure), budget availability, site access, slope steepness, and the availability of construction equipment and experienced contractors. The most commonly used method to repair surficial failures is to rebuild the failed area by pushing the failed soil mass back and re-compacting it.

Mechanical stabilisation techniques utilise rock, gabion baskets, concrete, geosynthetics and steel pins to reinforce slopes. These techniques can provide stability to both cut and fill slopes (Fay et al., 2012). Mechanical stabilisation techniques include retaining walls, mechanically stabilised earth, geosynthetically reinforced soil, and other *in situ* reinforcement techniques. For anchoring shallow soils, the use of *in situ* earth reinforcements and recycled plastic pins (RPPs) have been reported in slope stabilisation (Pearlman et al., 1992; Loehr et al., 2000).

Earthwork techniques involve the physical movement of soil, rock and/or vegetation for the purpose of erosion control and slope stabilisation. They involve reshaping the slope surface by methods such as creating terraces or benches, flattening oversteep slopes, soil roughening and land-forming. In addition, earthwork techniques can be used to control surface run-off, erosion and sedimentation during and after construction (Fay et al., 2012). The different techniques available in practice to stabilise surficial slope failure are presented in the following sections.

2.5.1 Slope rebuilding

Rebuilding the slope entails rebuilding the failed zone through compaction. This method consists of air-drying the failed soil, pushing it back to the failure area, and re-compacting it. Rebuilding the slope is considered one of the most economical methods of repair and is performed as routine maintenance work on failed slopes. However, this method is not effective for most scenarios, particularly in clays, as the shear strength of the soil is usually at the residual state, and compaction at the field level does not significantly increase the shear strength, especially when the soil becomes wet again. Consequently, repeated slope failures are observed for this method.

2.5.2 Pipe piles and wood lagging

This repair method consists of installing pipe piles and a wood lagging system in the failed zone, which provides resistance along the failed soil mass. During the process, the failed debris of the site is disposed of in different places, followed by cutting benches into the natural ground below the slip surface. Galvanised steel pipe piles are then

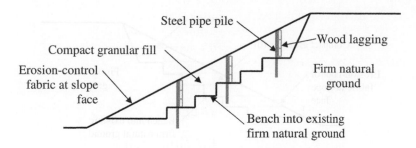

Figure 2.15 Schematic of pipe piles and wood lagging repair (Day, 1996).

installed (driven or placed in pre-drilled holes) and filled with concrete. Wood lagging (pressure-treated) is placed behind the piles, and a drainage system is then built behind the wood. A selected fill is compacted in layers, and the face of the slope is protected with erosion-control fabric and landscaping (Day, 1996). The schematic of pipe piles and wood lagging is presented in Figure 2.15.

One of the disadvantages of this method is that the lateral soil pressure against the wood lagging is transferred directly to the pipe piles, which are small in diameter and have low flexural capacity and low resistance to lateral loads. Pile failure by bending is a common occurrence in this repair method (Titi and Helwany, 2007).

2.5.3 Geosynthetic/geogrid repair

Geosynthetics/geogrids are fabricated from high-density polyethylene resins and are inserted inside the slopes as soil reinforcement. Reinforced soil slopes (RSSs) can generally be steeper than conventional unreinforced slopes, as geosynthetics provide tensile reinforcement that allows slopes to be stable at steeper inclinations. According to Elias et al. (2001), the design methods for RSSs are conservative, so they are more stable than flatter slopes designed to the same safety factor. RSSs offer several advantages over mechanically stabilised earth (MSE) walls. The backfill soil requirements for RSSs are usually less restrictive (the structure is more tolerant of differential settlement, and no facing element is required) which makes it less expensive compared to an MSE wall. Moreover, vegetation can be incorporated into the face of the slope for erosion protection.

Geogrids have an open structure which allows interlocking with granular materials used to rebuild slope failures. According to Day (1996), repair of surficial slope failures using geogrid materials consists of the complete removal of the failed soil mass. Benches are then excavated in the undisturbed soil below the slip surface. Vertical and horizontal drains are installed to collect water from the slope and dispose of it off-site. Finally, the slope is built by constructing layers of geogrid and compacted granular material. The schematic of geogrid repair is presented in Figure 2.16.

The repair of the slope using geosynthetics/geogrids requires excavation below the failure zone and may require an excavation support system during construction, which may drive the repair costs to a very high level.

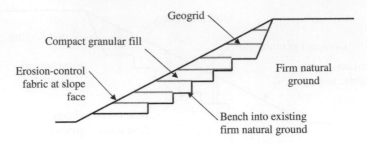

Figure 2.16 Repair of surficial slope failure by geogrid (Day, 1996).

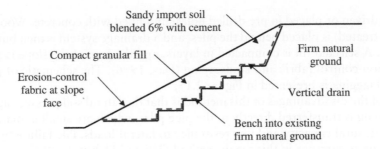

Figure 2.17 Soil-cement repair of shallow slope failure (Day, 1996).

2.5.4 Soil-cement repair

Soil-cement repair for shallow slope failures is similar to geogrid repair and is conducted by excavation and removal of the failed zone. Benches are then excavated in the undisturbed soil below the slip surface, and drains are installed to collect water from the slope and dispose of it off-site. Granular fill material is usually mixed with cement (~6%), and the mix is compacted to at least 90% of modified Proctor maximum unit weight (Day, 1996). The soil-cement mix will develop high shear strength and lead to a slope with a higher factor of safety. The schematic of soil-cement repair is presented in Figure 2.17.

The repair of the slope using this technique may require a temporary excavation support system during construction, which may drive up the repair costs. In addition, uniformly mixing the cement and treating the clay soil is challenging in field conditions, and this may cause the formation of some weak spots in the slope, leading to failure.

2.5.5 Repair with launched soil nails

In this method, soil nails are inserted into the slope face at high speed, using high-pressure compressed air. The soil nails are installed in a staggered pattern throughout the failed zone, which provides resistance along the slipping plane and increases the factor of safety, as presented in Figure 2.18. Typical soil nails can be solid or hollow steel bars; however, galvanised soil nails can also be used in highly abrasive environments,

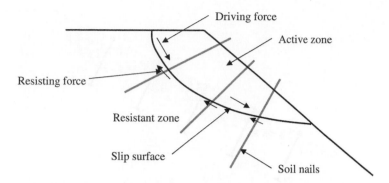

Figure 2.18 Schematic of repair of soil nails in slope stabilisation (replotted after Titi and Helwany, 2007).

as they provide resistance to corrosion. Typical hollow, non-galvanised steel bars have an outer diameter of 1.5 in (wall thickness of 0.12 in) with a length of 20 feet. The suggested minimum yield strength of the steel bars is 36 ksi (Titi and Helwany, 2007). After installing the launched soil nails, the slope surface can be treated with erosion-control matting, steel mesh or shotcrete.

The cost of a soil nail wall is determined by many factors, including type of ground, site accessibility, length of the nail, thickness of facing, type of construction (such as temporary or permanent), and availability of skilled manpower. In general, soil nail walls are 30% cheaper than tieback walls. A major cost item for a permanent soil nail wall is the wall facing. In many repairs for slope stabilisation in the North Texas area, a temporary vertical soil nail wall is constructed to retain the cut section and provide an uninterrupted flow of traffic. However, these temporary shoring walls contribute a major portion of the total project cost, which drives up the overall repair cost of this method.

The typical cost range for soil nail walls, based on US highway project bidding experience, ranges between $300 and $600 per square meter, where the costs are total in-place costs per square metre of wall face area.

2.5.6 Earth anchors

Earth anchors have been used in many geotechnical applications, including stabilisation of surficial slope failures. Earth anchoring systems consist of a mechanical earth anchor, wire rope/rod, and end plate with accessories. Repair of surficial slope failures with earth anchoring systems starts with viewing the failed slope. The earth anchors are installed by pushing the anchors into the ground below the failure surface. The wire tendon of the anchor is pulled to move the anchors to their full working position. The wire tendon is then locked against the end plastic cap (end plate), and the system is tightened. A schematic of earth anchors for stabilising shallow slope failure is presented in Figure 2.19. The earth anchor method is not successful in fine-grained soils in the presence of water, and its application is limited.

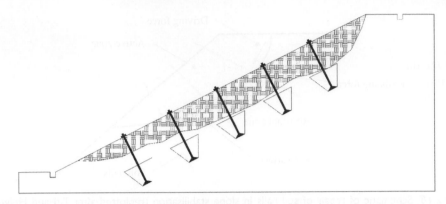

Figure 2.19 Earth anchors in slope stabilisation (redrawn after Titi and Helwany, 2007).

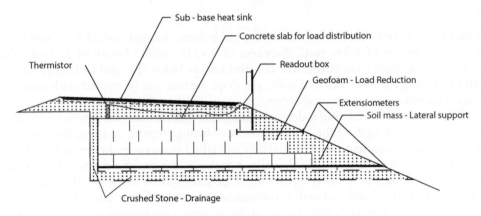

Figure 2.20 Typical section of geofoam stabilisation treatment.

2.5.7 Geofoam

Geofoam is a generic term for rigid cellular polystyrene, which is often used in geotechnical applications and has provided solutions worldwide for many difficult subsoils. The most common types of geofoam are expanded polystyrene (EPS) and extruded polystyrene (XPS). EPS is formed with low-density cellular plastic solids that have been expanded as lightweight, chemically stable, environmentally safe blocks. It generally behaves like an elastoplastic strain-hardening material. The unit weights of the material range from 0.7 to 1.8 pcf and it has compression strengths ranging between 13 and 18 psi.

Geofoam has been used to repair problematic highway slopes. In this technique, the failed soil is excavated and reconstructed using geofoam. A typical section of this slope repair technique is illustrated in Figure 2.20. Geofoam has a considerably lower unit weight than soil; however, it has high compressive strength. As a result, the driving force from the soil decreases, which eventually increases the factor of safety of the slope.

The field performance of this technique is promising. Jutkofsky et al. (2000) repaired a highway slope in New York using geofoam. The performance monitoring results showed that no post-construction lateral movement had taken place. In addition, the extensiometer reported negligible movement between the geofoam after construction. Another benefit is that it serves as an insulating material and resists temperature variations due to icing.

One disadvantage of this technique is that the failed soil must be replaced, which may require the installation of a temporary soil-retaining system for construction and repair of the failed slope. As a result, this method may be cost-prohibitive.

2.5.8 Wick drains

Santi et al. (2001) evaluated a new installation of horizontal geosynthetic wick drains to determine an effective option for stabilising landslides by reducing the amount of water that the slope contains. Horizontal wick drains are inexpensive, resist clogging, and may be deformed without rupture, thereby offering several advantages over conventional horizontal drains. A study was conducted by Santi et al. (2001) in which 100 drains were installed at eight sites in Missouri, Colorado and Indiana, using bulldozers, backhoes and standard wick drain-driving cranes. The study indicated that the drains were driven 30 m (98.4 ft) into the soil, with standard penetration test (SPT) values as high as 28. In addition, both experience and research indicate that drains should be installed in clusters that fan outward, aiming for an average spacing of 8 m (26.3 ft) for typical clayey soils.

Santi et al. (2001) installed and tested the effectiveness of horizontal wick drains during 1998 in an instrumented embankment in Rolla, Missouri. The embankment, which had a slope ratio of 1:1, was instrumented with six piezometers, 16 nested soil moisture gauges, and 20 survey markers. Half of the slope was stabilised using six wick drains, while the other half was retained as a control section. Behaviour of the wick drain was tested by artificial simulation of rainfall using a sprinkler. The result indicated that the wick drain removed a substantial amount of water from the slope, thereby lowering the groundwater level by 0.3 m and resulting in significantly less movement in the stabilised zone. The authors stabilised several test sites with varying geology, using various driving equipment. No evidence of clogging by dirt or algae was observed after installation, and the stabilisation scheme's performance was improved.

Based on this experience, Santi et al. (2001) suggested that the drains should not extend more than 3–5 m (10–15 ft) beyond the existing or potential failure surface. In addition, drains should be installed horizontally in clusters that fan outward, with eight-metre spacing. During installation, a smear zone was created that reduced the flow of water. The smear zone was reduced by pushing the pipe that contained the drain, rather than using pounding or vibration methods. The wick drains, however, have a few limitations. For the use of wick drains to be successful, the recommended SPT value must be 20 or less. The maximum drain length is expected to be 100 feet for harder soils and 150–200 feet for softer soils. In addition, there could be a significant number of dry drains on a project.

2.5.9 Retaining wall

Retaining structures are used to retain materials at a steep angle and are very useful when space (or right of way) is limited. Low-height retaining structures at the toe of a slope make it possible to grade the slope back to a more stable angle (flat slope) and can be successfully revegetated without loss of land at the crest. Such structures can also protect the toe against scour and prevent undermining of the cut slope. The advantage of using short structures at the top of a fill slope is that they can provide a more stable road bench or extra width to accommodate a road shoulder. Retaining structures can be built external to the slope (such as a concrete or masonry retaining wall), or can use reinforced soil (such as a burrito wall or deep patch) (Fay et al., 2012). It should be noted that the use of retaining walls in slope stabilisation can also be applied to large failures such as deep-seated failures, but this book is primarily focused on stabilising shallow slope failures. Examples of low-height retaining wall systems are presented in the following subsections.

2.5.9.1 Low masonry or concrete walls

Masonry or concrete retaining walls are rigid structures that do not tolerate differential settlement or movement and are appropriate only at sites where little additional movement is expected. In general, gravity walls can be constructed with plain concrete, stone masonry, or concrete with reinforcing bars. Masonry walls that incorporate mortar and stone are easier to construct and stronger than drystone masonry walls. However, they do not drain as well (Fay et al., 2012). Cantilevered walls use reinforced concrete and have a stem connected to a base slab.

A schematic of a low cantilevered retaining wall used to flatten a slope and establish vegetation is illustrated in Figure 2.21. Retaining walls with free-draining compacted backfill can be designed and constructed more efficiently than cohesive backfill soils. In this case, a drainage system should be installed behind the wall to facilitate the flow of water and inhibit the formation of a perched water zone behind the wall (Fay et al., 2012).

The low-height retaining wall has been proven to be effective and is a well-accepted method in the industry. However, stabilisation using the low-height retaining system is cost-prohibitive in some cases. Moreover, in fine-grained soil (especially in problematic expansive soil), failure of the wall is common due to the lack of sliding resistance. In addition, the use of fine-grained soil as a backfill creates a perched water zone, which also increases the lateral pressure on the slope and limits the performance of the drainage system. A photo of a cracked retaining wall over Highway Loop 12 in Dallas, Texas, is presented in Figure 2.22.

2.5.9.2 Gabion walls

Gabion baskets are made of heavy wire mesh and are assembled on site, set in place, then filled with rocks. Once the rocks have been placed inside the gabion basket, horizontal and vertical wire support ties are used to achieve the anticipated strength. Gabion walls are composed of stacked gabion baskets and are considered unbound structures. Their strength comes from the mechanical interlock between the stones or rocks (Fay et al., 2012). Gabion walls can be used at the toe of a cut slope or the top of a fill slope. The walls can be vertical or stepped and are adaptable to a wide range of slope geometries. Gabion walls can accommodate settlement without rupture and provide free drainage through the wall (Kandaris, 2007).

Cross Section

Not to scale

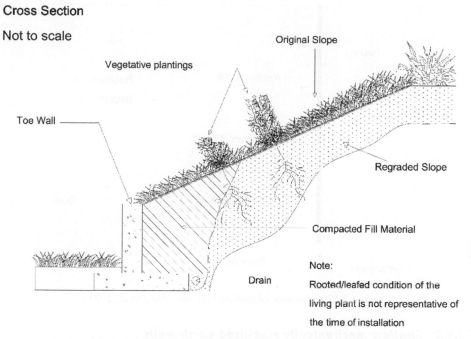

Original Slope

Vegetative plantings

Toe Wall

Regraded Slope

Compacted Fill Material

Drain

Note:

Rooted/leafed condition of the

living plant is not representative of

the time of installation

Figure 2.21 Cross section of a low wall with vegetation planted on the slope for stabilisation (USDA, 1992).

Figure 2.22 Failure of a low wall over highway Loop 12 in Dallas, Texas, due to sliding movement.

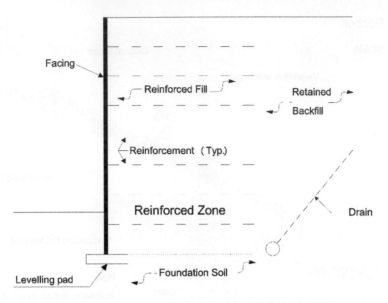

Figure 2.23 Schematic of shallow MSE wall (Berg et al., 2009).

2.5.9.3 Shallow mechanically stabilised earth walls

MSE walls are constructed with reinforced soil, as illustrated in Figure 2.23. The reinforcement can consist of metal strips (galvanised or epoxy-coated steel), welded wire steel grids, or geogrids. MSE walls can be designed and built to accommodate complex geometries to heights greater than 80 feet. They offer several advantages over gravity and cantilevered concrete retaining walls, such as simpler and faster construction, less site preparation, lower cost, more tolerance for differential settlement, and reduced right-of-way acquisition (Elias et al., 2001).

The economic savings of MSE walls compared with traditional concrete retaining walls, are significantly greater at heights of more than ten feet; however, lower MSE walls can also be constructed economically (Fay et al., 2012). For shallow MSE walls, the least expensive option is usually modular block facing, rather than precast concrete or metal sheet (Elias et al., 2001). It is recommended that good-quality backfill materials be used to facilitate drainage, especially for high walls; low walls can, however, be constructed using poorer quality soils (Fay et al., 2012).

The MSE system is a popular choice due to its cost-effectiveness. However, in soft clay foundation soil, the wall has less sliding resistance and may undergo recurring failure.

2.5.10 Pin piles (micropiles)

Pin piles (also known as micropiles) are more commonly used for foundations than slope stabilisation (Tarquinio and Pearlman, 1999). However, they have great potential for use in slope stabilisation but have been used only in very limited applications (Fay et al., 2012), primarily for deep-seated slope failures.

2.5.11 Slender piles

Flexible and rigid piles have been used recently for slope stabilisation applications. The free field soil movements associated with slope instability induce lateral load distributions along structural elements, which vary with the p–y response (where p is the soil–pile reaction and y is the pile deflection), pile stiffness, and the section capacity of the piles. In this case, each pile element offers passive resistance to lateral soil movement by transferring the loads to a stable foundation. Basically, there are two approaches described in the literature: the pressure-based method and the displacement-based method. The stabilising piles are designed as passive piles in the pressure-based method, where ultimate soil pressures are estimated and applied to the piles directly or as an equivalent loading condition. The assumptions of the pressure-based method are often not satisfied for free-headed slender pile elements in cases of larger pile deformation or plastic flow. As an alternative, the pile soil reaction and passive pile response can be evaluated as a function of the relative displacement between the soil and piles. However, evaluation of the relative displacement between the soil and pile is complicated, as the pile displacement depends on the soil displacement near the pile; therefore, the analysis of the displacement response considers the soil–pile interaction.

2.5.12 Plate piles

Recently, plate piles have been used to stabilise shallow slope failures in California. The plate piles increase the resistance to sliding by reducing the shear stress, and are installed vertically into the slope, similar to the pile-slope system. In a typical application, the plate piles are 6–6.5 feet long, with steel angle iron sections of 2.5 × 2.5 in, and a 1 × 2 feet rectangular steel plate welded to one end (McCormick and Short, 2006). The plate piles are driven into an existing landslide or potentially unstable slope which has two to three feet of soil or degraded clay fill over stiffer bedrock, as shown in Figure 2.24.

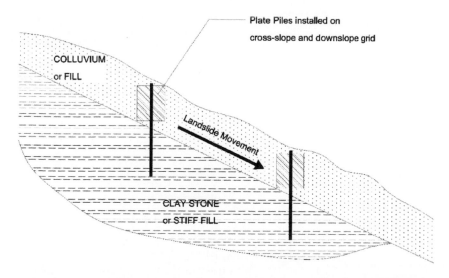

Figure 2.24 Schematic of plate piles for slope stabilisation (Short and Collins, 2006).

As a result, the plate reduces the driving forces of the upper slope mass by transferring the load to the stiffer subsurface strata too.

The critical component in determining the initial pile spacing is the angle iron resistance (Short and Collins, 2006). In an experimental test section, the plate piles were installed on a staggered grid pattern at four-feet intervals, centre to centre. Depending on the stiffness of the underlying materials, plate piles can be installed by the direct-push method, using an excavator bucket, or with a hoe-ram or 'head-shaker' compactor, at rates of 20–25 blows per hour.

This shallow slope stabilisation method, using plate piles, is the latest innovation and offers much potential as an alternative approach. The field implementation and controlled slope experiments conducted by Short and Collins (2006) showed that the plate pile technique can increase the factor of safety against slide by 20% or more, and can reduce the cost of slope stabilisation by six to ten times that of conventional slope repairs. The major demerits of the technique include a limited failure depth that only ranges up to three feet.

2.5.13   Recycled plastic pins

Recycled plastic pins (RPPs) have been used in Missouri, Iowa and Texas as a more cost-effective solution for slope stabilisation than conventional techniques (Loehr and Bowders, 2007; Khan et al., 2015). Typically, RPPs are fabricated from recycled plastics and waste materials, such as polymers, sawdust and fly ash (Chen et al., 2007). They are made of a lightweight material and are less susceptible to chemical and biological degradation than other structural materials. RPPs are installed in failed areas to provide resistance along the slipping plane and increase the factor of safety. RPPs have significant financial and environmental benefits for stabilising shallow slope failures, and further details are presented in the remaining chapters of this book.

Chapter 3

Generation and recycling of plastics

3.1 INTRODUCTION

The use of plastic has increased significantly over the past few decades due to its relatively low production cost, high durability and light weight. Plastics are usually synthetic or semi-synthetic organic polymers that have a high molecular mass and are produced from the chemical reaction of petrochemicals formed from fossil fuels. Approximately 4% of annual petroleum production is converted directly into plastics from petrochemical feedstock. The plastics industry has advanced over the years as different methods for the production of polymers from petrochemical sources have been invented (Andrady and Neal, 2009; Thompson et al., 2009). In 2007, world polymer production, including thermoplastics, thermoset plastics, adhesives and coatings, was estimated to be 260 million metric tons per annum (PlasticsEurope, 2008). Due to its light weight, plastic consumes less energy (i.e. fossil fuel) in transportation than similar volumes of steel (Andrady and Neal, 2009; Thompson et al., 2009). Plastic has a wide variety of applications, from short-term use (e.g. packaging, agricultural industries and disposable consumer items) to long-term infrastructure materials (e.g. pipes, cable coatings and structural materials). Plastic can also be used for durable consumer applications with an intermediate lifespan, such as electronic goods, furniture and vehicle parts. As a consequence of the ever-growing number of plastic products and the wide variety of their applications in modern daily life, post-consumer plastic waste generation has also been increasing rapidly (PlasticsEurope, 2008).

3.2 GENERATION OF PLASTIC WASTE

3.2.1 Global scene

Globally, approximately 1.3 billion tons of waste are currently generated each year. By 2025, this number is projected to be 2.2 billion tons. Figure 3.1 shows the average waste composition of the world, where organic waste comprises the majority of the waste, followed by paper, metal, plastic and glass. An average of 10% of the main waste stream is plastic, which varies according to the income level of the country. In low-income countries, the plastic content is found to be as low as 8%, whereas in high-income countries it comprises up to 11% of the total waste stream (Hoornweg and Bhada-Tata, 2012).

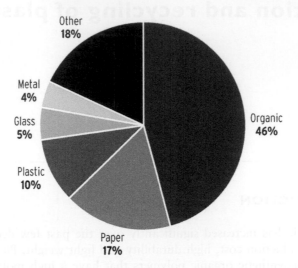

Figure 3.1   Global municipal solid waste composition in 2009 (Hoornweg and Bhada-Tata, 2012).

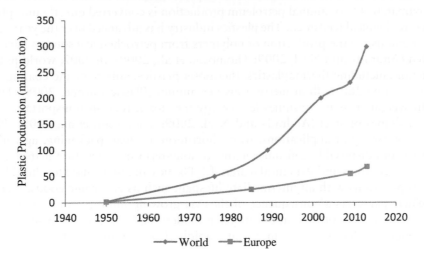

Figure 3.2   Global production of plastic products (reproduced from PlasticsEurope, 2013).

Global plastic production has experienced an exponential growth trend over the last 60 years. Figure 3.2 shows that 299 million tons of plastics were produced worldwide in 2013, compared to 1.5 million tons in 1950 (PlasticsEurope, 2013). Figure 3.3 shows plastic production by region in 2013. According to PlasticsEurope (2013), China's production was the highest (24.8%), followed by Europe (22.9%), the Middle East and Africa (20.8%), and North America (19.4%). The amount of plastics produced over the past ten years is almost equal to all of the plastics produced over the previous 30 years. This increase in the manufacturing of plastic products also increases the worldwide generation of plastic waste. According to the United Nations Environment Programme (UNEP, 2014), between 22% and 43% of the plastics used

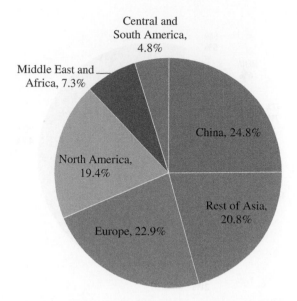

Figure 3.3 World production of plastic materials by region (reproduced from PlasticsEurope, 2013).

worldwide are disposed of in landfills. Plastics, being non-degradable in nature, stay in the landfill eternally, effectively wasting valuable landfill space. Figure 3.4 depicts the plastic waste generation levels in different parts of the world.

3.2.2 US perspective

In the United States, about 254 million tons of waste were generated in 2013. Of that, plastic waste constituted 32.52 million tons, making it 12.8% of the total municipal solid waste (MSW) (USEPA, 2015), as shown in Figure 3.5. A wide variety of plastics is present in the plastic waste stream (Figure 3.6), including polyethylene terephthalate (PET), which constitutes 14.4% (USEPA, 2015). The number of PET bottles has been increasing in the waste stream, with 4.68 million tons generated in 2013 (USEPA, 2015), compared to 0.32 million tons in 1997 (Subramanian, 2000). The volume of plastic consumption has been increasing steadily in the USA. Plastics made up an estimated 390,000 tons of the MSW generated in 1960, and increased relatively steadily to 32.5 million tons in 2013. As a percentage of MSW generation, plastics comprised less than 1% in 1960, but had increased to 12.8% by 2013.

3.3 MANAGEMENT OF PLASTIC WASTE

3.3.1 Global scene

Recycling is one of the most preferred options in the waste management hierarchy for reducing environmental impact. Plastics recycling began in the 1970s, as plastic's non-degradability came into focus. In some countries, the recycling of plastic materials has expanded rapidly in subsequent decades. New technological advances in collecting,

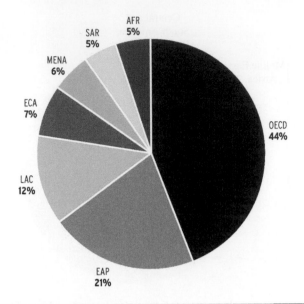

AFR: Africa Region
EAP: East Asia and Pacific Region
ECA: Europe and Central Asia Region
LAC: Latin America and the Caribbean Region
MENA: Middle East and North Africa Region
OECD: Organisation for Economic Co-operation and Development
SAR: South Asia Region

Figure 3.4 Plastic waste generation by regions (Hoornweg and Bhada-Tata, 2012).

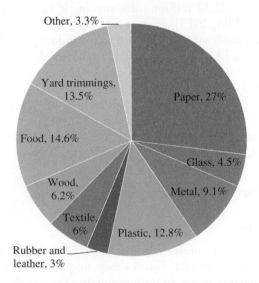

Figure 3.5 Composition of MSW in USA in 2013 (reproduced from USEPA, 2015).

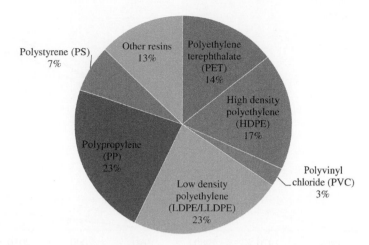

Figure 3.6 Types of plastics in MSW in USA in 2013 (reproduced from USEPA, 2015).

sorting and reprocessing recyclable plastics are diverting a major fraction of plastic waste from landfills. The quantities of recycled plastic vary geographically according to plastic type and application.

With the increasing global production of plastic, there is also rapid generation of plastic waste: about 150 million tons per year (Singh et al., 2016). The three most-used techniques for the management of plastic waste are depositing into landfills, incineration for energy recovery, and recycling (Bassi, 2017). Globally, according to the United Nations Environmental Programme (UNEP, 2014), between 22% and 43% of used plastic is disposed of in landfills, while in developed countries up to 21% of plastic waste is sent to incineration plants to generate energy. Collection rates of plastic waste, however, are poor in some places; for example, 57% of plastic in Africa, 40% in Asia, and 32% in Latin America remains uncollected. In 2007, the annual global trade of waste plastics was 15 Mt, and it was estimated that it would reach 45 Mt by 2015. By 2020, the worldwide demand for recovered plastics is expected to be 85 Mt (Plastics-Europe, 2015). In 2014, 25.8 million tons of post-consumer plastic waste ended up in upstream waste in Europe, of which 29.7% was recycled, 39.5% was incinerated for energy generation, and 30.8% went to landfills. Plastic waste recycling in Europe increased from 4.7 million tons in 2006 to 7.7 million tons in 2014, a 64% increase in recycling and a 38% reduction in the use of landfills (PlasticsEurope, 2015). Some countries in Europe are achieving higher recycling rates and have already banned the use of landfill. However, although waste management practices have evolved substantially over recent decades with an increasing trend towards recycling, landfill is still the dominant waste management practice in Europe and the rest of the world.

3.3.2 US perspective

In the United States, approximately 32.5 million tons of plastic waste were generated in 2013, but only 9% was recycled. The remaining 29.5 million tons were discarded

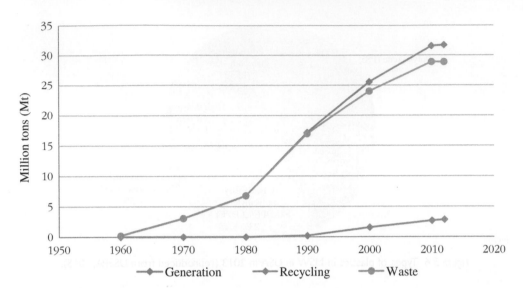

Figure 3.7 Plastic generation, recovery and waste in USA from 1960 to 2013 (reproduced from USEPA, 2015).

into landfills (USEPA, 2015). Figure 3.7 shows the generation, recovery and landfilling of plastic waste in the USA from 1960 to 2013. The figure shows that the levels of plastic generation and plastic waste generation were almost identical until recycling started in the 1990s. However, even in 2013, the recycling rate remains minimal, and almost all of the plastic generated in the USA ends up in landfills as waste.

Not all types of plastic are recyclable. Thermoplastics, such as PET, PE and PP, have high potential for being mechanically recycled (Rebeiz and Craft, 1995). Among these, PET seems to have the highest consumption and recovery rates due to its wide variety of applications in the packaging industry. Around the world, the annual consumption of PET for beverage bottles is more than 10 million tons, and its production and consumption is still increasing by 10–19% every year (Song et al., 2011; Zhang et al., 2006). In the US, 4.68 million tons of PET waste was generated in 2013, with 0.93 million tons recovered (19.9%) and the remaining 80.1% discarded (Figure 3.8). Among this mass of PET waste, PET bottles (soft drink and water bottles) accounted for 2.88 million tons, with a recovery rate of 31.3%. Recovery of natural (uncoloured) HDPE bottles (e.g. milk bottles and water jugs) was estimated at 220,000 tons, or 28.2% of generation, whereas the total HDPE waste generation was 5.58 million tons with just 10.2% recovery.

3.3.3 Potential benefits of recycling plastic waste

Worldwide, the rate of plastic recycling is still minimal, with a maximum of 30% in Europe and only 10% in the USA (PlasticsEurope, 2013). The remaining plastic that is not recycled/recovered is discarded into landfill.

Unlike most other components of the total waste stream, plastic and plastic products do not degrade over time. Although plastics are lightweight, they occupy a large

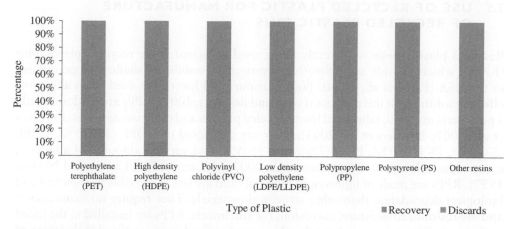

Figure 3.8 Discards and recovery of different types of plastic waste in USA (reproduced from USEPA, 2015).

volume of landfill space and remain in the landfill forever, reducing the amount of space available for more environmentally friendly waste burial. Therefore, the diversion and/or reuse of plastic products from landfill could save landfill space and help the landfill to stabilise at a faster rate in the long run. The diversion of plastics from landfill also creates a market for recycled plastics.

3.4 USE OF RECYCLED PLASTICS IN DIFFERENT APPLICATIONS

The use of recycled plastics enhances sustainable resource management by reducing the environmental impact of using virgin material. Recycled plastic has extensive uses in many applications and products, some of which are described below.

Construction: Recycled plastic has been widely used in mainstream construction products, such as damp-proof membranes, drainage pipes, ducting and flooring, due to its durability, low weight, low maintenance, resistance to vandalism, and non-degradability. Recycled plastic lumber is also used for pier construction. Another notable use of recycled waste plastic is the production of recycled plastic pins, which are used primarily to stabilise highway slopes.

Packaging: Recycled PET and HDPE are widely used in primary packaging by retailers and brand manufacturers for plastic bottles and trays.

Apparel: Textile fibre for clothing, such as polyester fleece, and polyester filling for duvets, coats, etc. is made from recycled PET bottles, and is the largest market for recycled PET bottles worldwide.

Landscaping and street furniture: Recycled plastic has been used increasingly to build walkways, jetties, pontoons, bridges, fences and signs. Traffic management products, such as street signs, street furniture, seating, bins and planters, are also being made from plastic.

Bag, sack and bin liners: New film products, such as carrier bags, refuse sacks and bin liners, are made on a large scale from old plastic film from sources such as pallet wraps, carrier bags and agricultural films.

3.5 USE OF RECYCLED PLASTIC FOR MANUFACTURE OF RECYCLED PLASTIC PINS

Recycled plastic waste is currently being used to manufacture recycled plastic pins (RPPs), which provide an innovative approach to stabilising shallow slope failures in the USA (Parra et al., 2004). Both Missouri and Iowa have used them as a cost-effective solution for this purpose (Loehr and Bowders, 2007). RPPs are predominantly a polymeric material, fabricated from recycled plastics and other waste materials (Chen et al., 2007; Bowders et al., 2003). They are composed of HDPE (55–70%), LDPE (5–10%), PS (2–10%), PP (2–7%), PET (1–5%) and varying amounts of additives, e.g. sawdust, fly ash (0–5%) (Chen et al., 2007; McLaren, 1995; Lampo and Nosker, 1997). RPPs are made of lightweight materials and are less susceptible to chemical and biological degradation than other structural materials. They require no maintenance and are resistant to moisture, corrosion, rot and insects. RPPs are installed in the failed area to provide resistance along the slipping plane and increase the slope's factor of safety. RPPs have great potential to become a popular and cost-effective alternative for the stabilisation of shallow slope failures.

The use of RPPs made from recycled plastic bottles for highway slope stabilisation and civil engineering infrastructure projects represents a perfect example of sustainable engineering solutions. Because plastic is not degradable, it is problematic if discarded into landfills. However, it can be very useful and beneficial in civil engineering infrastructure projects because soil slopes and highway soil slopes repaired with RPPs retain their engineering characteristics for a long time, thereby reducing the overall maintenance and repair cost of slopes and pavement shoulders in the long run.

Chapter 4

Recycled plastic pins

4.1 INTRODUCTION

The recycled plastic pin (RPP), which is commercially known as recycled plastic lumber, is manufactured by using post-consumer waste plastic and has been recommended as an acceptable material for use in the construction of docks, piers, bulkheads, etc. Plastic lumber is also marketed as an environmentally preferred material. From an environmental and life cycle cost analysis (LCCA) standpoint, RPPs are under serious consideration for use as structural materials for marine and waterfront applications. RPPs require no maintenance and are resistant to moisture, corrosion, rot and insects. The use of RPPs helps to reduce the problems associated with the disposal of plastics. Typically, 50% or more of the feedstock used for plastic lumber is composed of polyolefins, in terms of high-density polyethylene (HDPE), low-density polyethylene (LDPE) and polypropylene (PP). The polyolefins act as an adhesive and combine high-melt plastics and additives, such as fibreglass and wood fibres, within a rigid structure. Further additives, including pigments, foaming agents and ultraviolet (UV) stabilisers, are added during the process of manufacturing recycled plastic lumber.

4.2 MANUFACTURING PROCESS OF RPPs

The manufacturing of recycled plastic lumber begins with the collection of the raw material, which is plastic waste. After collection, the plastic is cleaned and pulverised, and the resulting confetti is delivered to the production site. Malcolm (1995) presented two methods of manufacturing recycled plastic lumber: an *injection moulding* process and a *continuous extrusion* process. In the injection moulding process, molten plastic is injected into a mould that defines the shape and length of the product. The mould is then cooled uniformly, and the product is removed from the mould. The process is relatively simple and inexpensive; however, the production volume is limited (Malcolm, 1995).

The continuous extrusion process allows production of varying lengths of the recycled plastic lumber. During this process, molten plastic is continuously extruded through a series of dies which shape the material during its cooling. However, it is challenging for the manufacturer to provide uniformly controlled cooling of the sample to prevent warpage and caving of the lumber.

The continuous extrusion process requires a considerable upfront investment compared to the injection moulding process. However, it requires less labour and is suitable for rapid mass production.

Compression moulding is another widely used process for the manufacture of recycled plastic goods (Lampo and Nosker, 1997). This process melts batches of thermoplastics, typically 50–70% of the mix, with other materials. An automatically adjusted scraper then removes the molten material from the plasticator and presses it through a heated extruder die into premeasured, roll-shaped loaves. The loaves are then conveyed to a press-charging device that fills a sequence of compression moulds. The products are cooled in the moulds to a temperature of 40°C and ejected onto a conveyor which moves them to a storage area.

4.3 ENGINEERING PROPERTIES OF RPPs

RPPs are manufactured by using commingled waste plastic that has been collected by municipal recycling programmes. Typically, 50% or more of the feedstock used for plastic lumber manufacturing consists of HDPE, LDPE and PP. Since the RPPs are made from recycled plastics, the composition of the virgin materials varies within the recycled plastic, resulting in differences in the composition of the RPPs produced.

In slope stabilisation, RPPs are utilised to increase lateral resistance, and to increase the factor of safety of the slope and resist the movement of the soil. In addition, during the installation process to stabilise the slope, RPPs must have resilience to the impact loading while being driven into the ground. As a result, the compressive and tensile strength, as well as the flexural resistance, of RPPs is of interest to geotechnical engineers. The potential modes of failure of RPPs in slope stabilisation are illustrated in Figure 4.1.

The elastic modulus of an RPP plays a major role in controlling the deformation of surficial soil. Several studies have indicated that plastics are susceptible to damage due to UV radiation and deformation (creep) under sustained loading. When RPPs are installed in the ground, they are subjected to pH levels in the soil which may cause deterioration of the structural capacity of the RPPs. Therefore, the following sections consider the structural capacities and important parameters that affect the performance of RPPs.

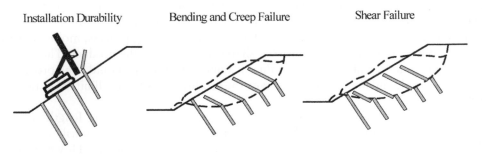

Installation Durability   Bending and Creep Failure   Shear Failure

Figure 4.1 Potential failure modes of RPPs in slope stabilisation applications (Loehr and Bowders, 2007).

4.3.1   Compressive and tensile strength

Bowders et al. (2003) conducted a study on the different engineering properties of RPPs. The motivation for the study was the evaluation of the engineering properties of a wide variety of production standards and the development of specifications appropriate to slope stabilisation. Samples were collected from three different manufacturers and uniaxial compression tests were performed. The results of these tests are presented in Table 4.1.

Lampo and Nosker (1997) conducted a comparative experimental study on the compressive strength of recycled plastic lumber. During the study, ten plastic samples were obtained from eight manufacturers. The composition of the products varied greatly. Some were mixed plastics, some were pure resins, and others contained fillers, such as wood pulp or fibreglass. Lampo and Nosker (1997) performed the experimental study according to The American Society for Testing and Materials (ASTM) specification D695-85, with a sample height of 12 inches. To calculate the mechanical properties, the study included an effective cross-sectional area which was calculated on the basis of a specific gravity measurement. The compressive strength test was performed at a loading rate of 0.1 in/min. Based on the results, the modulus, ultimate strength at 10% strain, and yield strength at 2% offset were calculated from the load-displacement data. The specific modulus and specific strength are the modulus divided by specific gravity, and the ultimate strength divided by specific gravity, respectively. These "specific" properties represent the mechanical properties of the materials, normalised with respect to density, during the study. It was assumed that this normalisation would minimise the effects of voids when comparing the materials' properties, and the effects of extrusion by different methods during the manufacturing process, which varies among manufacturers. The compressive strength results are presented in Table 4.2.

Based on their experimental study, Lampo and Nosker (1997) reported that the compression values for RPP lumber ranged between 12.0 MPa (1,740 psi) and

Table 4.1 Uniaxial compression test results (Bowders et al., 2003).

Specimen batch	No. of specimens tested	Nominal strain rate (%/min)	Uniaxial compressive strength (MPa)		Young's modulus, $E_{1\%}$ (MPa)		Young's modulus, $E_{5\%}$ (MPa)	
			Avg.	Std. dev.	Avg.	Std. dev.	Avg.	Std. dev.
A1	10	–	19	0.9	922	53	390	27
A2	7	0.005	20	0.8	1285	69	378	15
A3	6	0.006	20	0.9	1220	108	363	27
A4	3	0.004	20	0.9	1377	165	363	25
A5	4	0.006	12	1	645	159	225	17
A6	4	0.006	13	0.9	786	106	238	34
B7	2	0.007	14	0.5	541	36	268	3
B8	2	0.006	16	0.4	643	1	308	0.5
C9	3	0.0085	17	1.1	533	84	387	40

Note: 1 MPa = 145 psi.

Table 4.2 Average values of various strength-related parameters for different samples of recycled plastic lumber (Lampo and Nosker, 1997).

Sample	Specific gravity	Modulus (MPa)	Specific modulus (MPa)	Yield stress (MPa)	Ultimate strength (MPa)	Specific strength (MPa cm³/g)
51A	0.2789	262	840	4.89	5.41	19.4
1B	0.7012	427	609	9.52	13.0	18.6
2D (BR)	0.8630	588	682	11.5	16.0	18.5
2D (G)	0.8098	800	988	14.5	19.7	24.3
1E	0.862	557	647	12.2	16.7	19.4
1F	0.7888	746	945	15.1	19.4	24.6
1J (B)	0.7534	643	854	13.1	16.3	21.6
1J (W)	0.9087	759	836	14.9	19.5	21.4
23L	0.7856	1320	1680	11.8	13.3	16.9
1M	0.5652	399	705	6.65	8.45	15.0
1S	0.9090	555	610	11.5	14.1	15.5
1T	0.8804	813	921	15.5	21.5	24.4
9U	0.774	598	769	12.6	16.6	21.3

Note: 1 MPa = 145 psi.

24.1 MPa (3,500 psi); the tension values ranged from 8.62 MPa (1,250 psi) to 24.1 MPa (3,500 psi). However, the RPPs reached their ultimate strength at different strain levels.

Breslin et al. (1998) compared the different test results presented in the literature, as shown in Table 4.3. The study reported that the incorporation of different additives (glass fibre, wood fibre, polystyrene) into plastic lumbers increased the stiffness of the lumber.

Plastic is a temperature-dependent material. At low temperatures, plastic is strong and brittle; however, with an increase in temperature, the plastic becomes weaker and more ductile. Malcolm (1995) reported the effects of temperature change on the tensile strength of HDPE materials, as presented in Figure 4.2.

4.3.2 Flexural strength

Ahmed (2013) conducted a study on the flexural capacity of a structural grade RPP (FIBERFORCE® from Bedford technologies). A total of nine RPP samples were tested to evaluate their flexural strength, using the three-point bending test in accordance with ASTM D790. The test was performed at the structural engineering laboratory of the University of Texas at Arlington (UTA). Loehr and Bowders (2007) concluded that the strength and stiffness of RPPs are sensitive to the loading rate. Bowders et al. (2003) conducted an experimental study of RPPs for slope stabilisation, where the deformation rates utilised were five to ten times lower than the suggested value in ASTM D6109. Because the loading rate in slope stabilisation is much lower than the loading rate suggested in ASTM D790, Ahmed (2013) conducted the three-point bending test at three different loading rates (0.5, 2.7 and 4.9 kip/min) that fell between the range recommended by ASTM D790 and the loading rate suggested for slope stabilisation. Three samples were tested for each loading rate and the stress–strain responses at different loading rates are presented in Figure 4.3. The flexural strength

Table 4.3 Engineering properties of plastic lumber products (Breslin et al., 1998).

Product	Composition	Compressive strength (psi)	Modulus of elasticity (psi)	Tensile strength (psi)	Source
TRIMAX	HDPE/glass fibre	1740	450000	1250	TRIMAX literature
	HDPE/glass fibre			1189	SUNY at Stony Brook
Hammer's Plastic Recycling	80% HDPE/20% LDPE	89814			Zarillo and Lockert (1993)
	HDPE/LDPE (20PSGF)	527000			
	HDPE/LDPE (40PS20GF)	653000 (D790)	1793 (D638)		
Superwood Selma, Alabama	33% HDPE/33% PP	3468 (D695)	146171 (D790)	1793 (D638)	Zarillo and Lockert (1993)
California Recycling Company	100% commingled	81717			
	10% PP	79319			
	50% HDPE	92636 (D790)			Beck (1993)
RPL-A	HDPE/glass fibre	2000			Smith and Kyanka (1994)
RPL-B	49% HDPE/51% wood fibre				Smith and Kyanka (1994)
Rutgers University	100% curb tailings	3049	89500		Renfree et al. (1989)
	60% milk bottles, 15% detergent bottles, 15% curb tailings, 10% LDPE	3921	114800		Renfree et al. (1989)
	50% milk bottles, 50% densified polystyrene	4120 (D695)	164000 (D790)		Renfree et al. (1989)
BTW recycled plastic lumber	Post-consumer	1840–2801	162000		BTW/ Hammer's brochure

and elastic modulus of the RPPs ranged from 3.1 to 4.7 ksi, and from 190 to 200 ksi, respectively. The experimental results were subsequently used in the design of slope remediation.

Bowders et al. (2003) also investigated the variations in the flexural strength of RPPs. RPP samples were collected from three different manufacturers, and four-point bending tests were performed. The results are presented in Table 4.4.

4.3.3 Effect of weathering on long-term properties

Krishnaswamy and Francini (2000) performed a study on the effect of outdoor weathering and environmental effects, including degradation due to UV radiation, thermal expansion, and the combined effects of moisture and temperature on the mechanical behaviour of RPPs. The authors did not observe significant variations of the flexural

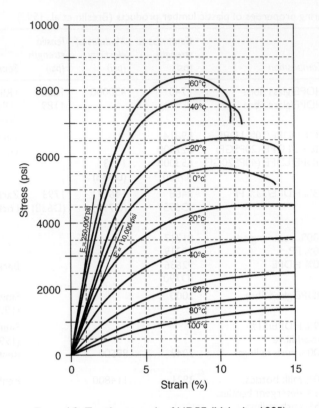

Figure 4.2 Tensile strength of HDPE (Malcolm, 1995).

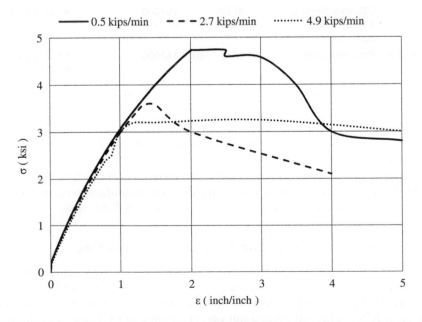

Figure 4.3 Stress–strain response of an RPP at different loading rates (three-point bending test).

Table 4.4 Four-point bending test results (Bowders et al., 2003).

Specimen batch	No. of specimens tested	Nominal deformation rate (mm/min)	Flexural strength (MPa)	Secant flexural modulus, $E_{1\%}$ (MPa)	Secant flexural modulus, $E_{5\%}$ (MPa)
A1	13	–	11	779	662
A4	3	4.27	18	1388	–
A5	3	5.74	11	711	504
A6	4	3.62	10	634	443
B7	1	4.05	9	544	425
B8	1	5.67	–	816	–
C9	2	3.21	12	691	553

Note: I MPa = 145 psi.

Table 4.5 Comparison of flexural properties of typical RPP materials with and without hygrothermal cycling (Krishnaswamy and Francini, 2000).

	Secant modulus (psi)	Stress at 3% strain (psi)
Before cycling	97800 ± 6400	1900 ± 120
After cycling	113600 ± 14400	2400 ± 400

modulus or the strength of the RPPs, relative to ASTM D6109, before or after the hygrothermal cycling, as documented in Table 4.5.

Lynch et al. (2001) investigated the effect of weathering on the mechanical behaviour of recycled HDPE-based plastic lumber. During the study, the flexural properties of weathered deck boards were obtained by performing flexural tests in three-point loading, for comparison to original flexural properties, according to ASTM D796. The study investigated both the exposed and unexposed sides of the deck boards in tension. The original properties of the RPP deck boards, determined before the weathering action, indicated a flexural modulus of 1,179 MPa (171,000 psi) and a flexural strength of 17.24 MPa (2,500 psi). The three-point bending test results of the weathered samples when the exposed and unexposed sides were tested in tension are presented in Tables 4.6 and 4.7, respectively.

Comparing these test results with the original mechanical properties, it was observed that both the modulus and strength increased after the outdoor exposure. The modulus increased by 28% from the original when the exposed side was tested in tension and by 25% when the unexposed side was tested in tension.

Breslin et al. (1998) conducted a study on the long-term engineering properties of plastic lumber manufactured by using post-consumer plastic. During the study, the plastic lumber samples were removed from the deck over a two-year period and returned to the laboratory for testing. At the outset, the authors investigated the initial engineering properties of the recycled plastic lumber, which was manufactured using a continuous extrusion process. The plastic lumber samples were collected at regular intervals during the two-year monitoring period. The lumber did not face severe traffic; however, it was subjected to two summer cycles of high temperatures and UV intensity.

Table 4.6 Three-point bending test results of RPP samples after weathering (exposed side tested in tension) (Lynch et al., 2001).

Sample	Modulus (MPa)	Strength at 3% strain (MPa)	Ultimate strength (MPa)
1A	1658	19.07	23.68
2A	1474	17.13	21.52
3A	1385	16.82	19.69
4A	1477	17.61	22.89
5A	1568	18.81	22.84
Average	1512	17.89	22.12

Note: 1 MPa = 145 psi.

Table 4.7 Three-point bending test results of RPP samples after weathering (unexposed side tested in tension) (Lynch et al., 2001).

Sample	Modulus (MPa)	Strength at 3% strain (MPa)	Ultimate strength (MPa)
1B	1500	19.10	24.03
2B	1410	17.03	21.01
3B	1312	16.87	21.04
4B	1512	16.78	21.41
5B	1618	19.03	22.39
Average	1470	17.76	21.98

Note: 1 MPa = 145 psi.

The authors did not observe any noticeable change, such as warping, cracking and/or discoloration of the plastic lumber.

Breslin et al. (1998) investigated the compression modulus of the sample periodically, both in cross-sectional and in-plane axes, as presented in Figure 4.4. The study indicated that the measured in-plane compression modulus (192 MPa or 27,850 psi) was six to eight times higher than the cross-sectional compression modulus (24 MPa or 3,480 psi). In addition, no significant change was observed in either of the compression moduli over a period of 19 months; however, the 24-month modulus was significantly higher in both of the planes when compared to the initial value. The significant change over the 24-month period may have been the result of variability in the material properties of the lumber profiles.

Breslin et al. (1998) also presented a study on the modulus of elasticity based on different compositions of plastic lumber. The study included the modulus of elasticity in the in-plane direction, as well as in the cross-sectional direction. Based on the study, the variations of modulus of elasticity of plastic lumbers from different manufacturers are summarised in Figure 4.5.

Breslin et al. (1998) observed that the poorest engineering properties were found in lumber manufactured using a mixture of post-consumer waste plastics. The use of a single polymer, such as HDPE with glass fibre additives, results in significantly better engineering properties. In addition, the use of glass and wood fibre additives significantly improves the modulus of elasticity for plastic lumber. The modulus of

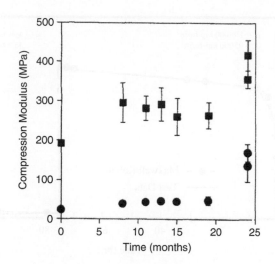

Figure 4.4 Compression modulus measurements for both cross-sectional and in-plane dimensions for plastic lumber collected from the West Meadow Pier over a 24-month period (Breslin et al., 1998).

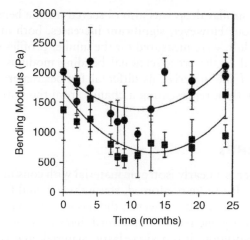

Figure 4.5 Bending modulus measurements for both cross-sectional and in-plane directions for plastic lumber collected from the West Meadow Pier over a 24-month period (Breslin et al., 1998).

elasticity for plastic lumber manufactured without fibre additives ranged from 79 to 173 ksi (545–1,193 MPa); with the addition of wood or plastic fibres, the modulus of elasticity ranged from 146 to 653 ksi (1,007–4,502 MPa) which is two to four times higher than the lumber without fibre additives.

Breslin et al. (1998) observed a higher initial bending modulus in the cross-sectional direction (1,860 Pa) when compared to the in-plane direction (1,400 Pa). The authors also observed changes in the bending modulus over time for the samples collected from the pier, as presented in Figure 4.5. The authors noticed a significant decrease in the bending modulus measured in the in-plane direction (750 Pa compared to an initial

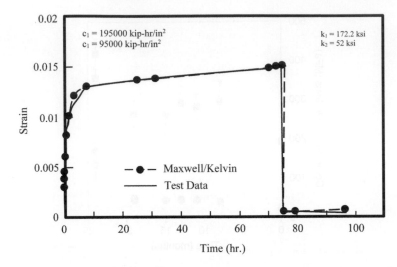

Figure 4.6 Creep curve for recycled plastic beam at room temperature (Malcolm, 1995).

value of 1,370 Pa). A similar drop was also observed for the bending modulus in the cross-sectional direction. However, significant increases, both in in-plane and cross-sectional bending moduli, were measured for the lumber profiles tested at 24 months. The changes measured in the cross-sectional bending modulus may be a reflection of the heterogeneity of the material, as different lumber materials were collected at different times, rather than representing a change from the initial lumber properties due to weathering.

4.3.4 Creep of RPPs

Recycled plastic lumber is a nearly isotropic material with considerable strength, durability and workability. It can be reinforced, strengthened, and formed as a composite material. It is as strong as wood. However, the modulus of elasticity of unreinforced plastic lumber is generally one-tenth to one-third that of southern yellow pine lumber (Malcolm, 1995). In addition, it is a viscoelastic material, and is susceptible to creep and increased deflection with time under a static sustained load.

Malcolm (1995) conducted a study on the creep behaviour of a $1.5\,\text{in} \times 3.5\,\text{in}$ recycled plastic lumber sample subjected to a sustained mid-span bending stress of 516.7 psi, which produced the creep curve presented in Figure 4.6. During this study, low stress level for the sustained load was maintained for plastic lumber.

Van Ness et al. (1997) conducted a study on the long-term creep behaviour of commercially available plastic lumber. Creep in recycled plastic lumber represents the time dependence of the mechanical properties of plastic lumber. To investigate the long-term creep behaviour, the study included four groups of plastic lumbers, manufactured by four different companies. The composition of the four plastic groups varied significantly. Some contained blends of polyolefin, and one contained glass fibres, but all of the samples were constituted principally of recycled polyethylene. Based on the experimental study, Van Ness et al. (1997) observed that the recycled plastic lumber that contained oriented glass fibre was the most creep-resistant over time.

Gopu and Seals (1999) performed a study on the effects on the mechanical properties of recycled plastic lumbers of composition, member size, service temperature, service stress level, and the duration and orientation of loading. The study indicated a clear lack of homogeneity in the plastic lumber, which influenced the orientation of loading with respect to member axes. Furthermore, the authors proposed that the flexural strength and stiffness values of the lumber be adjusted to account for the effect of creep and temperature. In addition, the authors emphasised the use of glass fibres to improve the stiffness of the recycled plastic lumber.

Lampo and Nosker (1997) reported that creep is a serious concern when using RPPs for any load-bearing application. Due to the viscoelastic properties of plastics, a piece of plastic lumber will begin to sag over time under a static load. This time-dependent effect increases with elevated temperature. Civil engineers study such time-dependent phenomena and develop load-duration factors for design use, and it is crucial that this effect is taken into account when developing design guidelines for plastic lumber.

4.3.4.1   Creep of RPPs in slope stabilisation

Chen et al. (2007) performed a study on the creep behaviour of RPPs. Due to the variety of manufacturing processes and constituents, the engineering properties of commercially available materials vary substantially. The polymeric materials are durable in terms of environmental degradation; however, they can exhibit higher creep rates than other structural materials, such as timber, concrete or steel.

Chen et al. (2007) tested eight 90 mm (3.5 in) × 90 mm (3.5 in) square specimens from three different manufacturers to evaluate creep behaviour. The *compressive creep* tests were performed on specimens cut from full-size RPPs with same sectional measurement of 90 mm (3.5 in) × 90 mm (3.5 in) × with 180 mm (7.2 in) in length. The compressive load was applied using a spring with a 44.1 kN/m (390,320 lb-in) spring constant. All specimens were tested at a temperature of 21°C. The *flexural creep* responses were performed on scaled RPPs of 3 cm × 3 cm × 61 cm (1.2 in × 1.2 in × 24 in), and the flexural creep test was performed at different temperatures (21, 35, 56, 68 and 80°C). The test setups for both compressive and flexural creep are presented in Figure 4.7. The study applied the Arrhenius method to estimate the long-term creep behaviour.

A typical plot of deflection vs. time for the compressive creep, and the results of a typical compressive creep test are presented in Figure 4.8 and Table 4.8, respectively. For all specimens, the primary creep was completed within one day of the load being applied. Secondary creep occurred after the primary creep and continued for a year at a steady rate.

The flexural creep test results are presented in Table 4.9. As the temperature increased, the time to reach failure decreased for the same load condition. The results showed that the loading levels, along with the temperature, affected the creep behaviour of the recycled plastic specimens. In addition, the higher the load levels or the closer to the ultimate strength of the material, the faster the creep rate and the shorter the time to reach failure. Based on the study, the authors presented a method

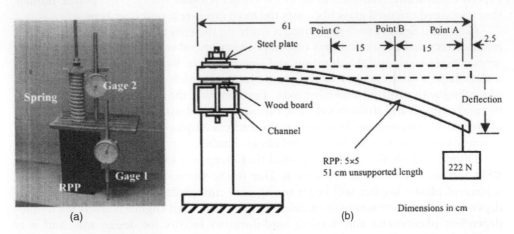

Figure 4.7 RPP testing setup for creep: a. compressive creep; b. flexural creep (Chen et al., 2007).

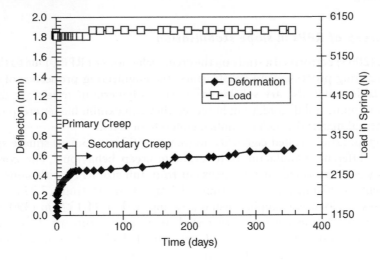

Figure 4.8 Typical deflections of RPP under constant axial stress vs. time (Chen et al., 2007).

for the investigation of the design life of RPPs, based on percentage load mobilisation, as shown in Figure 4.9. The higher the mobilised loads, the more susceptible to creep failure within its design life the RPP became. The authors suggested the performance of an effective design exercise to reduce the load mobilisation, which could be achieved by increasing the number of RPPs (thereby reducing the spacing), changing the constituents, or changing the section of the RPPs to increase their moment of inertia.

4.4 EFFECT OF ENVIRONMENTAL CONDITIONS

Ahmed (2013) conducted a study at UTA to determine the axial compressive strength of RPPs under different extreme environmental conditions. The uniaxial compressive

Table 4.8 Summary results of typical compressive creep test (Chen et al., 2007).

Manufacturing	No. of specimens	Creep stress (kPa)	Ratio of creep stress to compressive strength	Maximum creep strain (%)
A3	2	724	3.7	0.1
A6	2	690	6.3	0.1
B7	1	758	5.3	0.4
C9	1	827	5.1	0.4

Table 4.9 Summary of flexural creep test results (Chen et al., 2007).

Loading conditions	Temperature (°C)	No. of specimens tested	Average time to reach failure (days)	Comments
44 N at five points	21	2	1.185[a]	Not failed
	56	2	195	Failed
	68	2	3.5	Failed
	80	2	0.8	Failed
93 N single load	21	2	1.185[a]	Not failed
	56	2	574	Failed
	68	2	17.5	Failed
	80	2	8.5	Failed
156 N single load	21	2	1.185[a]	Not failed
	56	2	71.5	Failed
	68	2	0.6	Failed
	80	2	0.8	Failed
222 N single load	21	2	1.185[a]	Not failed
	35	4	200	Failed
	56	2	3.1	Failed
	68	2	0.4	Failed
	80	2	0.8	Failed

[a]Last day of testing: specimen not ruptured.

strength test was performed on RPPs in accordance with ASTM D6108. Both the normal samples and samples submerged in environmental chambers for two months were tested. Recycled plastic lumber reinforced with fibre glass was utilised, with a cross-sectional size of 3.5 in × 3.5 in. The test samples were prepared with the specimen height being twice the minimum width, in accordance with the ASTM standards. As 3.5-inch samples were utilised for all of the tests, the height of each sample was seven inches.

Ahmed (2013) evaluated the environmental effects in relation to acidic (pH = 5.5), neutral (pH = 7.0) and basic (pH = 8.5) conditions, which represent the pHs of different clayey soils in Texas. The samples were submerged in three large tubs, filled either with water, a basic solution or an acidic solution, for two months to study the degradation behaviour at an accelerated rate. For acidic and basic solutions, the samples were kept in sealed, covered containers inside the laboratory at room temperature

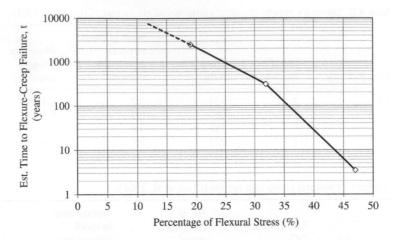

Figure 4.9 Method to estimate flexural creep in the field (Chen et al., 2007).

Table 4.10 Compression tests on RPPs in different environmental conditions.

Uniaxial compression test	Loading rates	No. of tests
Normal conditions	2.5 kip/min 3.1 kip/min 3.75 kip/min	3
Acidic solution, pH 5.5		3
Neutral solution, pH 7.0		3
Alkaline solution, pH 8.5		3

Figure 4.10 Uniaxial compression test of normal and weathered samples.

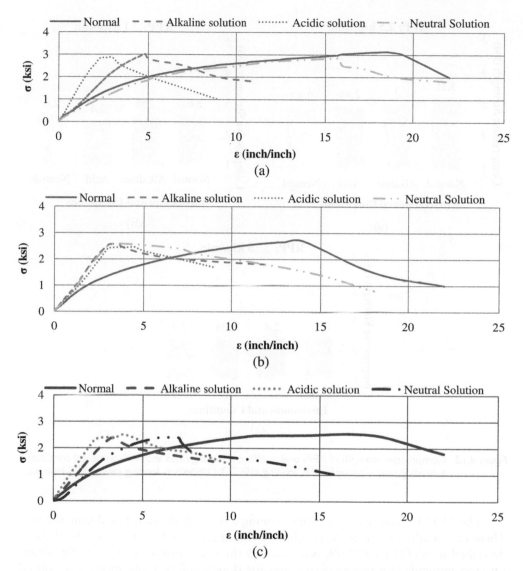

Figure 4.11 Stress–strain response of RPPs under different loading rates and environmental conditions; a. loading rate = 2.5 kip/min; b. loading rate = 3.1 kip/min; c. loading rate = 3.75 kip/min.

(70°F). The pH of the solution was measured each week to monitor the variations of solution concentration. In addition, the pH was adjusted, to keep it in the target range, by adding an acidic and/or basic solution if any variation was observed. The samples submerged in neutral solutions (pH = 7.0) were kept open in a hot room, with no covers, to simulate the effect of heat and moisture on the samples. The temperature of the hot room was maintained at 98°F, resembling the average temperature during summers in Texas.

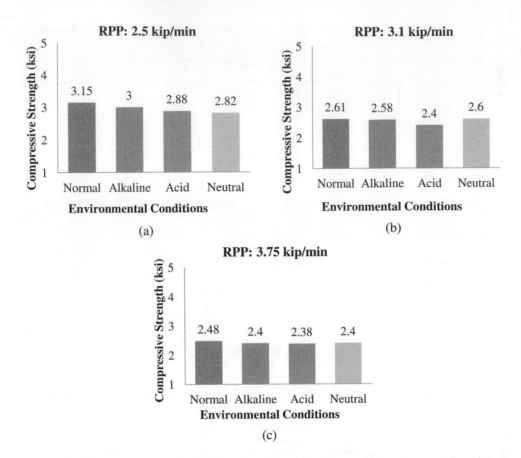

Figure 4.12 Compressive strength of RPPs under different loading rates and environmental conditions;
a. loading rate = 2.5 kip/min; b. loading rate = 3.1 kip/min; c. loading rate = 3.75 kip/min.

The ASTM standards suggest maintaining a controlled strain rate during the test. Therefore, loading rates, corresponding to the strain rates for the plastic lumbers as described in ASTM D6108-09, were used as the upper limits of the test. The strain rate recommended for testing plastic pins for slope stability applications was utilised as the lowest strain rate. A third strain rate was chosen in-between these upper and lower values. Three samples were tested for each loading rate. The experimental programme developed for this study is outlined in Table 4.10. The axial compression tests performed on the specimens with normal and environmental exposures are illustrated in Figure 4.10.

The stress–strain behaviour of RPPs for loading rates of 2.5, 3.1 and 3.75 kip/min in different environmental conditions is represented in Figure 4.11. According to the results, the highest axial strength was observed at the lowest loading rate (2.5 kip/min) under all environmental conditions and for all of the samples tested. In addition, the peak strength decreased with increased loading rate, resulting in lower strength regardless of any environmental effect.

The elastic modulus of an RPP is considered to be the initial slope of its stress–strain curve. The stress–strain curve shifted to the left with the application of both acidic and basic conditions, signifying the increment of elastic modulus during axial compression. The stress–strain curve of the sample in neutral conditions also shifted to the left and followed a trend similar to both the acidic and basic conditions, except for the samples submerged in the neutral solution of pH 7.0 at a loading rate of 2.5 kip/min. However, regardless of the environmental conditions, no significant change in the peak strength of the RPPs was observed during the experimental period.

The average maximum compressive strength of the RPP samples for all of the specimens tested during this study are presented in Figures 4.11 and 4.12. The compressive strengths of the RPPs were constant, regardless of the environmental conditions, for all loading rates. It is important to note that RPPs are composite materials that mainly contain high-density polyethylene, HDPE (55–70%), which is a non-degradable material. As a result, RPPs experience hardly any degradation due to environmental exposure.

The non-degradation behaviour of RPPs offers potential benefits for slope stabilisation. When installed in the ground, RPPs are exposed to different levels of pH, which over time usually tend to weather/degrade other construction materials, such as steel, concrete or timber pile. However, as RPPs are resistant to differential pH variations, almost no degradation of their strength is expected over time, which will ensure a longer design life than other alternative construction materials.

Chapter 5

Design methods

5.1 DESIGN METHODS

RPPs are installed in the slope to intercept potential sliding surfaces, providing additional resistance to maintain the long-term stability of the slope. The definition of the factor of safety (FS) of a slope is the ratio of the resisting moment (MR) to the driving moment (MD), as presented in Equation 5.1. RPPs installed in the slope provide an additional resisting moment (ΔMR) along the slip surface, thereby increasing the resistance and the FS, as presented in Equation 5.2. A schematic diagram of RPPs as slope reinforcement is presented in Figure 5.1.

$$FS = MR/MD \tag{5.1}$$

$$FS = (MR + \Delta MR)/MD \tag{5.2}$$

where MR = resisting moment along slip surface, MD = driving moment along slip surface, and ΔMR = additional resisting moment from plastic pin.

The RPPs provide direct resistance along the slip surface, resulting in the additional resisting moment, as represented in Equation 5.2.

Several field studies have been performed during recent decades to evaluate the performance of RPP-stabilised slopes. Based on these studies, Loehr et al. (2004) developed a "limit state design method". For this technique, they referred to the direct resistance of an RPP as the limit resistance, which varies along the depth of the RPP. The authors considered two soil failure mechanisms and two structural failure modes, and established a limit resistance curve for RPPs. This limit state design method is simple and very straightforward. The design approach considers the failure modes for the structural failure of RPPs. However, the design does not consider deformation of the reinforced slope or the creep failure of the RPPs.

Taking into consideration the limitations of the previous design methods, another set of field studies has been performed in Texas during the past few years, resulting in a performance-based design technique. This design approach was developed with consideration of three criteria that limit the failure of adjacent soil, control the deformation of the slope, and resist the creep failure of RPPs.

The following section summarises the limit state design method. In addition, the development of the limiting criteria and design approaches of the performance-based method are presented in the subsequent sections.

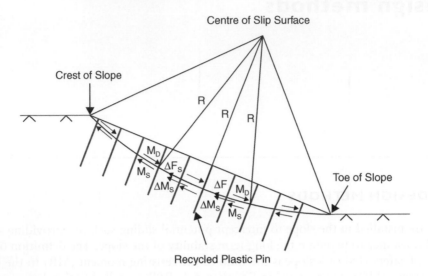

Figure 5.1 Schematic diagram of slope reinforcement with RPPs (Khan et al., 2016).

5.2   LIMIT STATE DESIGN METHOD

Loehr and Bowders (2007) presented a design approach using the limit equilibrium method to evaluate the stability of RPP-reinforced slopes. The general approach of the limit equilibrium method is to consider a potential sliding surface, then calculate its FS. The FS of a slope is considered as:

$$Factor\ of\ Safety\ (FS) = \frac{\int s}{\int \tau} \tag{5.3}$$

where s is the shear strength from soil at the sliding surface, and τ is the mobilised shear stress required to maintain equilibrium.

The slope stability analysis is conducted using the method of slice approach with the Mohr-Coulomb failures envelope, where the sliding body is divided into a number of vertical slices. The equilibrium of the individual slices determines the normal and shear stresses on the sliding surface, and the FS is determined on the assumption of a sliding surface. The process is repeated for other potential surfaces until the lowest FS value is obtained. For the reinforced slope, a procedure similar to the method of slice is used. A force, due to the reinforcing member, is added to the other forces on the slice that are intersected by reinforcing members. The schematic of the equilibrium of the unreinforced and reinforced slopes is presented in Figure 5.2.

The reinforcement force provides a direct resistance along the slip surface and increases the FS. Loehr and Bowders (2007) referred to the reinforcement force as the limit resistance, which varies along the depth of the RPP. The authors considered two soil failure mechanisms and two structural failure modes to establish the limit resistance curve for RPPs. Based on the failure modes of soil and RPPs, Loehr and Bowders (2007) proposed a combined limit resistance curve, as presented in Figure 5.3.

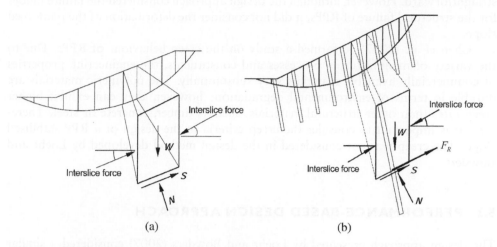

Figure 5.2 Static equilibrium of an individual slice in the method of slices: a. unreinforced slope; b. reinforced slope (Loehr and Bowders, 2007).

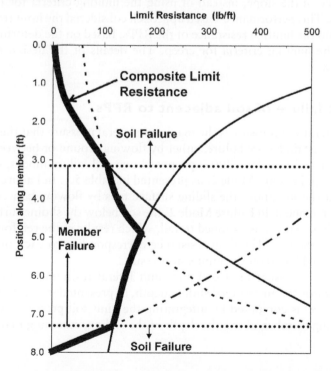

Figure 5.3 Combined limit resistance curve (Loehr and Bowders, 2007).

Once the combined limit resistance curve has been developed for a particular condition and slope, the FS of the reinforced slope can be determined by using limit resistance slope stability analysis software, including the resistance from the RPPs. The design approach considered by Loehr and Bowders (2007) was simple and very straightforward. However, although the design approach considered the failure modes for the structural failure of RPPs, it did not consider the deformation of the reinforced slope.

Chen et al. (2007) performed a study on the creep behaviour of RPPs. Due to the variety of manufacturing processes and constituents, the engineering properties of commercially available materials vary substantially. The polymeric materials are durable in terms of environmental degradation; however, they can exhibit higher creep rates than other structural materials, such as timber, concrete or steel. Therefore, it is important to consider the creep criteria in the design of a RPP-stabilised slope. The creep was not considered in the design method developed by Loehr and Bowders.

5.3 PERFORMANCE-BASED DESIGN APPROACH

The design approach presented by Loehr and Bowders (2007) considered a similar limit resistance approach to that used for the failure mode of soil. It also incorporated the performance of the slope, instead of using the limiting criteria for the structural failure of RPPs. This performance-based approach considered the limit resistance from adjacent soil and the limiting resistance of the RPPs, based on the deformation criteria of RPPs and the limiting criteria for creep. The details of the design approach are presented below.

5.3.1 Limit failure of soil adjacent to RPPs

The limiting lateral soil pressure is the maximum lateral pressure that the soil adjacent to the RPP can sustain before failure, either by flowing around or between reinforcing members. Loehr and Bowders (2007) proposed two soil failure modes, referred to as Failure Mode 1 and Failure Mode 2, as presented in Table 5.1. In Failure Mode 1, it is considered that the soil above the sliding surface fails by flowing between or around the reinforcing members. In Failure Mode 2, the soil below the sliding surface, adjacent to the reinforcing member, is assumed to fail, which results in the reinforcing member passing through the soil. The limit resistance corresponding to each of these failure modes is computed based on the limit soil pressure.

The limit soil pressure (a stress) and limit lateral resistance (a force) act on a reinforcing member for an assumed sliding depth, as presented in Figure 5.4. The limit resistance (a force) is computed by integrating the limit soil pressure over the length of the reinforcing member, above the depth of the sliding, assuming that the limit soil

Table 5.1 Summary of soil failure mode for establishing limit lateral resistance of RPPs (Loehr and Bowders, 2007).

Failure mode	Description
Failure Mode 1	Failure of soil above sliding surface or between reinforcing member
Failure Mode 2	Failure of soil below sliding surface due to insufficient anchorage length

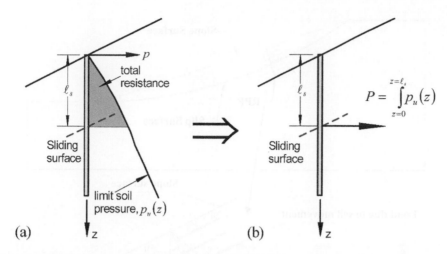

Figure 5.4 Schematic diagram of calculation of limit resistance force: a. limit soil pressure; b. equivalent lateral resistance force (Loehr and Bowders, 2007).

pressure is fully mobilised along the entire length of the member above the sliding surface.

5.3.2   Limit resistance of RPPs

The limit resistance of RPPs is evaluated using the performance criteria. Since the modulus of elasticity of RPPs is low compared to other structural materials, it is important to consider the anticipated displacement due to the applied soil pressure. In addition, the creep criteria should be considered in the limit resistance of the RPP, and the limit resistance should be evaluated based on the limit horizontal displacement, as well as the maximum flexural stress allowed on the RPP. The estimation of the limit resistance of an RPP is described below.

5.3.2.1   Limit horizontal displacement of RPPs

RPPs are subjected to active pressure and passive resistance of soil when installed as slope reinforcements. In addition, due to the sliding of the slope, they are also subjected to an additional soil pressure above the slip surface. The schematic of the load acting on the RPP is illustrated in Figure 5.5. RPPs get anchorage from the foundation soil below the slipping plane, work as a lateral support, and resist the movement of soil above the slip surface. However, during these interactions, a displacement takes place which depends on the additional pressure due to slope movement, the depth of the soft soil over the slipping plane, the active and passive pressure of the soil, and the anchorage from the foundation. The overall displacement of the slope is expected to take place based on the displacement of the RPP during this interaction. To control the overall displacement of the slope, it is important to limit the displacement of the RPPs acting as slope reinforcement. This design method considers the limit horizontal displacement approach, where the capacities of RPPs are evaluated based on the anticipated displacement due to the soil movement.

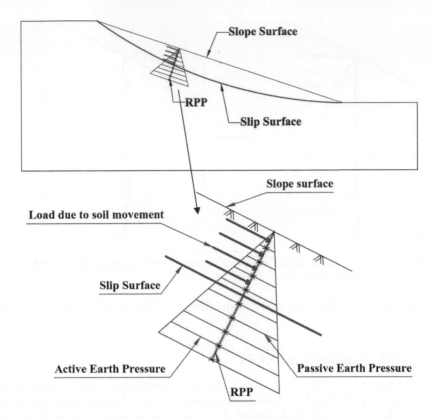

Figure 5.5 Schematic of load acting on an RPP used as slope reinforcement.

5.3.2.2 Limit maximum flexure for prolonged creep life

Chen et al. (2007) presented a study on the creep behaviour of RPPs. The authors studied the flexural creep responses of scaled RPPs of 1.2 in × 1.2 in × 24 in at different temperatures, such as 21, 35, 56, 68 and 80°C. The study utilised the Arrhenius method to estimate the long-term creep behaviour. Based on the study, Chen et al. (2007) observed that as the temperature increased, the time to reach failure decreased for the same load condition. In addition, the loading levels, along with temperature, affected the creep behaviour of the recycled plastic specimens. The study concluded that at higher load levels, closer to the ultimate strength of the material, the creep rate was faster and required a shorter time to reach failure. Based on the results of the study, the authors presented a method based on percentage load mobilisation to investigate the design life of RPPs, as shown in Figure 5.6.

Figure 5.6 indicates that at the higher mobilised loads, the design life of RPPs becomes susceptible to creep failure. Therefore, it is important to limit the percentage of flexural stress in RPPs, in order to increase the creep failure time. It was observed that at 35% of flexural stress, the estimated time to flexure-creep failure should be 100 years, which is longer than the average design life of a highway slope. Therefore, on the basis of this current design procedure, an RPP should be restricted to 35% of the flexural stress of its ultimate capacity.

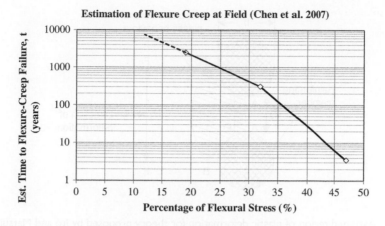

Figure 5.6 Flexural creep in field conditions (Chen et al., 2007).

5.4 DETERMINATION OF LIMIT SOIL PRESSURE

5.4.1 Calculation of limit soil pressure

Ito and Matsui (1975) proposed a method to determine the limit of lateral force on stabilising piles in a slope. The method was established based on soil failure between piles, assuming the soil between the piles to be in a plastic state, according to the Mohr–Coulomb failure criterion. The assumed region of plastic behaviour is shown in Figure 5.7. This method is referred to as the "theory of plastic deformation". The limit soil pressure of RPPs is calculated using Equations 5.4a and 5.4b. A sample soil parameter, as presented in Table 5.2, was utilised to determine the limit soil pressure for RPPs, which is illustrated in Figure 5.8.

$$P(z) = cD_1 \left(\frac{D_1}{D_2}\right)^{N_\varphi^{1/2}\tan\varphi + N_\varphi - 1} \left[\frac{1}{N_\varphi \tan\varphi} \left\{ \exp\left(\frac{D_1 - D_2}{D_2}\right) N_\varphi \tan\varphi \right. \right.$$

$$\left. * \tan\left(\frac{\pi}{8} + \frac{\varphi}{4}\right) - 2N_\varphi^{1/2}\tan\varphi - 1 \right\} + \frac{2\tan\varphi + 2N_\varphi^{1/2} + N_\varphi^{-1/2}}{N_\varphi^{1/2}\tan\varphi + N_\varphi - 1} \right]$$

$$- c\left\{ D_1 \frac{2\tan\varphi + 2N_\varphi^{1/2} + N_\varphi^{-1/2}}{N_\varphi^{1/2}\tan\varphi + N_\varphi - 1} - 2D_2 N_\varphi^{-1/2} \right\}$$

$$+ \frac{\gamma z}{N_\varphi} \left\{ D_1 \left(\frac{D_1}{D_2}\right)^{N_\varphi 1/2 \tan\varphi + N_\varphi - 1} * \exp\left(\frac{D_1 - D_2}{D_2}\right) N_\varphi \tan\varphi * \tan\left(\frac{\pi}{8} + \frac{\varphi}{4}\right) - D_2 \right\}$$

$$(5.4a)$$

where

$$N_\varphi = \tan^2\left(\frac{\pi}{4} + \frac{\varphi}{2}\right) \tag{5.4b}$$

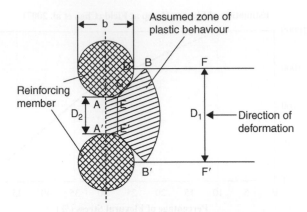

Figure 5.7 Assumed region of plastic deformation for theory proposed by Ito and Matsui (1975).

Table 5.2 Parameters for estimating limit soil pressure.

RPP properties		Soil parameters	
Size (inch) (rectangular)	3.5 × 3.5	c (psf)	200
Length (ft)	10	φ	10
Spacing (ft)	3	γ (pcf)	125

and c = cohesion intercept, φ = friction angle, γ = unit weight of soil, D_1 = centre-to-centre pile spacing, and D_2 = inner distance between piles.

5.4.2 Calculation of limit soil resistance

In Failure Mode 1, the soil above the sliding surface is assumed to fail by flowing between or around the reinforcing members. RPPs are assumed to be sufficiently anchored in stable soil below the sliding surface. The schematic diagram of Failure Mode 1 is presented in Figure 5.9. The limit resistance for Failure Mode 1 is determined by integrating the limit soil pressure (as presented in Figure 5.8) along the RPP, up to the depth of the sliding surface, considering that the limit soil pressure is fully mobilised along the length of the RPP below the sliding surface. This calculation is repeated for varying depths of sliding to develop a curve describing the magnitude of the limit resistance along the length of the RPP, as presented in Figure 5.10.

The limit soil resistance for Failure Mode 2 is calculated by a process similar to that for Failure Mode 1, except that the soil below the sliding surface adjacent to the RPP is assumed to fail, while the member is sufficiently anchored in the moving soil above the sliding surface. The reinforcing RPPs are essentially flowing through the soil below the sliding surface, as illustrated in Figure 5.11.

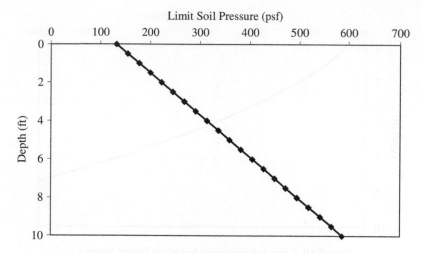

Figure 5.8 Limit soil pressure with depth of RPPs.

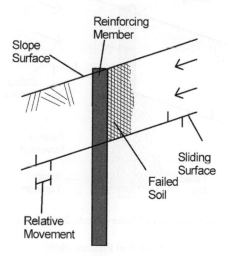

Figure 5.9 Schematic diagram of Failure Mode 1 (Loehr and Bowders, 2007).

The limit resistance for Failure Mode 2 is determined by integrating the limit soil pressure, as presented in Figure 5.8, along the RPP, below the depth of the sliding surface, assuming that the limit soil pressure is fully mobilised along the length of the member below the sliding surface. This calculation is repeated for varying depths of sliding to develop a curve that depicts the magnitude of the limit resistance along the length of the reinforcing member for Failure Mode 2, as presented in Figure 5.12.

By combining the two soil failure modes, a composite curve is developed by using the least resistance of the two failure modes to produce the limit resistance along the

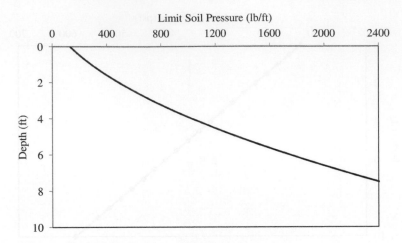

Figure 5.10 Limit soil resistance based on Failure Mode 1.

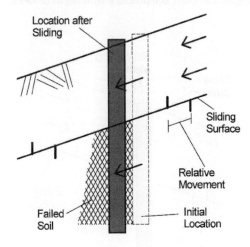

Figure 5.11 Schematic diagram of Failure Mode 2 (Loehr and Bowders, 2007).

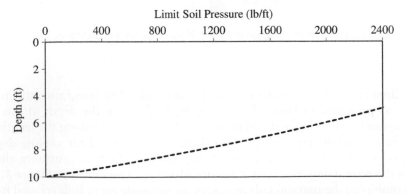

Figure 5.12 Limit soil resistance based on Failure Mode 2.

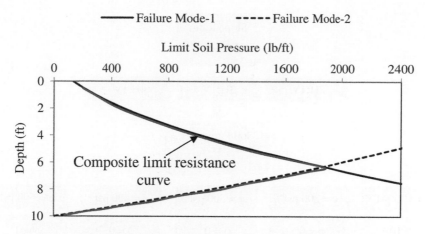

Figure 5.13 Composite limit resistance curve of Failure Modes 1 and 2.

Table 5.3 Considerations in the development of the design chart.

Slope inclination	Depth of slip surface (ft)	Lateral pressure on RPP (lb/ft²)
2H:1V, 3H:1V, 4H:1V	3, 4, 5, 6, 7	10, 20, 30, 40, 50, 60, 70, 80, 90, 100, 200, 300, 400, 500

Table 5.4 Soil parameters.

Cohesion (psf)		Friction angle φ		
100	0	10	20	30
200	0	10	20	30
300	0	10	20	30
400	0	10	20	30
500	0	10	20	30

length of the RPP, as presented in Figure 5.13. The limit resistance is suitable for cases where failure of the soil completely controls the resistance.

5.5 LIMIT HORIZONTAL DISPLACEMENT AND MAXIMUM FLEXURE OF RPPs

The elasto-plastic finite element method (FEM) is an accurate, robust and simple method. Previous studies indicated that the shear strength reduction in the FEM had

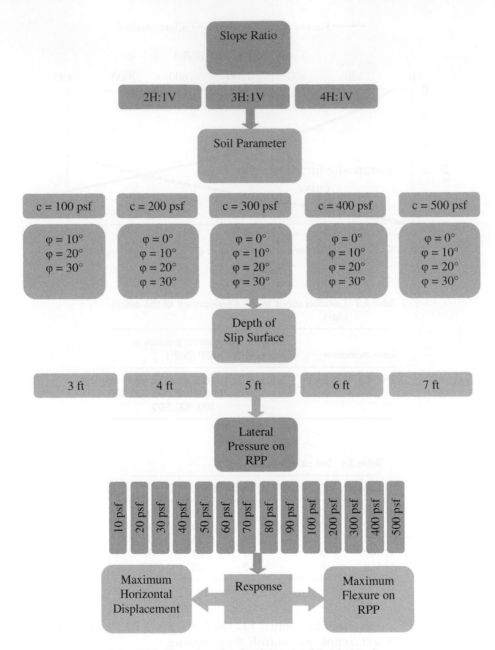

Figure 5.14 Flow chart for development of the design charts.

been successfully used to evaluate the stability and deformation of slopes reinforced with piles and anchors (Wei and Cheng, 2009; Yang et al., 2011; Cai and Ugai, 2003). This technique was utilised to evaluate the stability of slopes reinforced with piles or anchors under a general frame, where soil-structure interactions were considered,

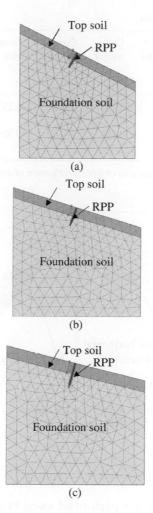

Figure 5.15 Soil model for determination of horizontal displacement with applied load: a. slope ratio 2H:1V; b. slope ratio 3H:1V; c. slope ratio 4H:1V.

using a zero-thickness elasto-plastic interface element. FEM was utilised to evaluate the resistance of RPPs.

The objective of the current approach was to develop a design chart to evaluate the load capacity of RPPs based on limiting horizontal displacement and maximum flexural stress. A series of loads was applied over the RPPs; the corresponding horizontal displacement as well as the maximum bending moment were determined for each case. RPPs were utilised to stabilise the shallow slope failure. Loehr and Bowders (2007) reported that the depth of the shallow failures ranged between 3 and 7 ft for moderate-to-steep slopes, where the geometry of the slope varied between 2H:1V and 4H:1V. It is also important to consider the soil strength, which should cover a wide range for shallow slope failures. This design method considers a wide range of failure depths, both loading and slope ratios, as well as soil strength parameters, as presented in

Table 5.5 Parameters from FE analysis.

Soil type —	Friction angle φ (°)	Cohesion c (psf)	Unit weight γ (pcf)	Elastic modulus E (psf)	Poisson ratio v —	RPP Property	Unit	Value
Top soil	10	200	125	2025	0.35	EA	lb/ft	857500
						EI	lbft²/ft	255300
Foundation soil	30	500	125	37140	0.30	d	ft	1.89
						w	lb/ft/ft	4.4

E = Modulus of elasticity, A = cross section area of RPPs, I = Moment of inertia, w = weight of RPP.

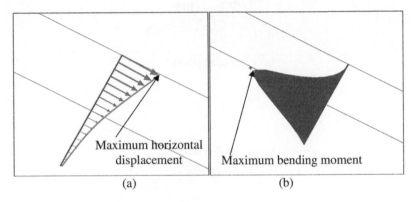

Maximum horizontal displacement

Maximum bending moment

(a)   (b)

Figure 5.16 Determination of response of RPP due to applied load: a. maximum horizontal displacement; b. maximum bending moment.

Tables 5.3 and 5.4. The flow chart for the development of the design chart is presented in Figure 5.14.

The deformation analysis was performed using PLAXIS 2D software. The RPPs were considered as elastic material and modelled as a plate element. The elastic, perfectly plastic Mohr-Coulomb soil model was utilised for stability analyses, using 15-node triangular elements. The 15-node element provides a fourth-order interpolation for displacements and numerical integration that involves 12 stress points. It is a very accurate element and has produced high-quality stress results for different problems (Plaxis, 2011). Standard fixities were applied as boundary conditions, where the two vertical boundaries were free to move vertically and were considered fixed in the horizontal direction. The bottom boundary was modelled as a fixed boundary.

The FEM analysis was performed using the soil model as presented in Figure 5.15. The FEM model was calibrated using the field monitoring data and performance of the slopes stabilised with RPPs. Once the calibration was done, the design chart was developed. It should be noted that because the bottom layer was considered as stiff foundation soil, two layers of soil were considered in the soil model for the design chart. The analysis was conducted with different soil strengths at the top soil, as presented earlier in Table 5.5. The deformation analysis was conducted by applying

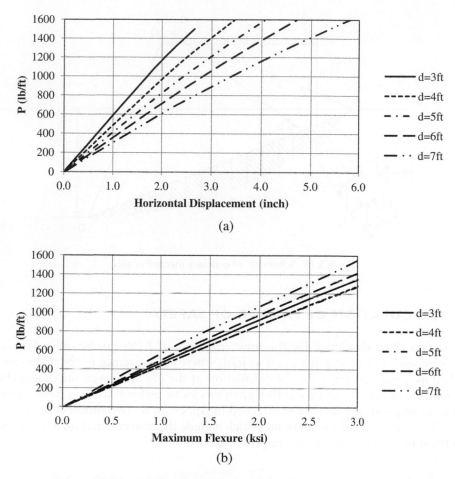

Figure 5.17 Limit resistance curves for RPP for $c = 200$ psf and $\varphi = 10°$: a. load vs horizontal displacement for slope 3H:1V; b. load vs maximum flexure for slope 3H:1V.

a uniform load over the RPPs throughout the sliding depth, and the corresponding maximum horizontal deformation. The maximum bending moment was also determined, as presented in Figure 5.16. Based on the applied load, the total resistance of the RPPs, with corresponding horizontal deformation and maximum flexure stress, is summarised for a given soil strength (for example, cohesion $c = 200$ psf, and friction angle $\varphi = 10°$) with varied depths of slip surface, as presented in Figure 5.17. Details of the parameters for the FEM analysis are presented in Table 5.5.

5.6 FINALISING THE DESIGN CHART

A series of design charts (limit soil failure, limit horizontal displacement and limit maximum flexure) were developed, based on different soil strength parameters. The design charts are illustrated in Appendix A.

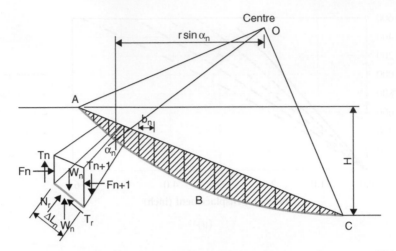

Figure 5.18 Schematic of ordinary method of slice.

5.7 CALCULATION OF FACTOR OF SAFETY

The design chart developed for the limit resistance of RPPs can be utilised, as can any commercial slope stability analysis tool, where the resistance of the RPPs can be applied as a pile resistance for the calculation of the FS. Moreover, the design chart can be utilised and the slope stability analysis can be performed using hand calculations. A sample design method is attached in Appendix B. Two different approaches to performing the hand calculations, which include the conventional method of slices and the infinite slope approach, are presented below.

5.7.1 Approach 1: conventional method of slices

The conventional method of slices, also known as the ordinary method of slices, can be explained using the schematic in Figure 5.18, where AC is a trial failure surface. Arc AC is centred at point O. The soil above the trial surface is divided into several vertical slices. Considering a unit-length perpendicular cross section below, the active forces that act on a typical slice are also presented in Figure 5.18, where W_n is the weight of the slice. The forces N_r and T_r, respectively, are the normal and tangential components of the reaction r. Pn and $Pn+1$ are the normal forces that act on the side of the slice. Similarly, the shearing forces that act on the sides are Tn and $Tn+1$. In applying the method of slices, the pore pressure is considered as zero to derive the equation. However, with the presence of water, the weight of the slice will be determined using the saturated unit weight, and the concepts of soil mechanics should be applied to the soil portion. It is important to note that the limit resistance of the RPP should be considered the same as for dry soil. Moreover, an approximation assumption is made in which Pn and Tn are equal in magnitude to the resultants of $Pn+1$ and $Tn+1$ and their lines of action coincide.

For the equilibrium condition,

$$N_r = W_n \cos \alpha_n \qquad (5.5)$$

and the resisting shear force,

$$T_r = \tau_d * b_n = \frac{\tau_f * \Delta L_n}{FS} = \frac{1}{FS}(c' + \sigma' \tan \varphi') * \Delta L_n \qquad (5.6)$$

the normal stress, σ', can be determined as:

$$\sigma' = \frac{N_r}{\Delta L_n} = \frac{W_n \cos \alpha_n}{\Delta L_n} \qquad (5.7)$$

For equilibrium of the trial wedge, ABC, the moment of the driving force about the centre point O equals the moment of the resisting force about O, as presented below:

$$\sum_{n=1}^{n=p} W_n * r * \sin \alpha_n = \sum_{n=1}^{n=p} \frac{1}{FS} \left(c' + \frac{W_n \cos \alpha_n}{\Delta L_n} \tan \varphi' \right) * \Delta L_n * r \qquad (5.8)$$

Therefore, the factor of safety of the unreinforced slope can be determined as:

$$FS = \frac{\sum_{n=1}^{n=p} (c' \Delta L_n + W_n \cos \alpha_n \tan \varphi')}{\sum_{n=1}^{n=p} W_n \sin \alpha_n} \qquad (5.9)$$

For the reinforced slope, RPPs provide an additional resistance, P, along the slipping plane and increase the resisting force, as presented in Figure 5.19. Therefore, for the reinforced slope:

$$T_r = \tau_d * b_n = \frac{\tau_f * \Delta L_n}{FS} = \frac{1}{FS} \left[\left(c' + \frac{W_n \cos \alpha_n}{\Delta L_n} \tan \varphi' \right) * \Delta L_n + P \right] \qquad (5.10)$$

For equilibrium of the trial wedge, ABC, the moment of the driving force about the centre point O, equals the moment of the resisting force about O, as presented below:

$$\sum_{n=1}^{n=p} W_n * r * \sin \alpha_n = \frac{1}{FS} \left[\sum_{n=1}^{n=p} \left(c' + \frac{W_n \cos \alpha_n}{\Delta L_n} \tan \varphi' \right) * \Delta L_n * r + \sum_{n=1}^{n=q} P * r \right] \qquad (5.11)$$

The FS of the reinforced slope can be determined as:

$$FS = \frac{\sum_{n=1}^{n=p} (c' \Delta L_n + W_n \cos \alpha_n \tan \varphi') + \sum_{n=1}^{n=q} P}{\sum_{n=1}^{n=p} W_n \sin \alpha_n} \qquad (5.12)$$

The composite soil resistance curve can be utilised to determine the limit resistance, P, of the RPPs, as presented in Appendix A.

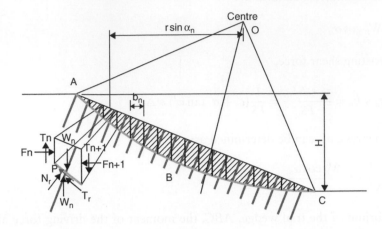

Figure 5.19 Schematic of reinforced slope using ordinary method of slice.

5.7.1.1   Design steps for approach 1

Most of the commercially available slope stability analysis programs can be utilised to determine the critical slip surface with minimum FS when designing a slope using RPPs. Once the critical slip surface is obtained, the reinforcement design can be performed by implementing the following steps:

- The failure surface of the slope should be divided into small segments of slices according to the conventional method of slice approach.
- A spacing of RPPs should be selected and plotted along the critical slip surface.
- Based on the plot of RPPs, the depth of the critical slip surface, *d,* that crosses the individual RPP can be determined.
- Using the depth of the critical slip surface, *d*, the limit resistance, *P*, of an individual RPP should be determined from the design charts, attached in Appendix A. The design chart should be utilised to determine the lowest resistance based on the limit soil resistance, limit horizontal displacement and maximum flexure criteria, taking into account the soil parameters and structural parameters of the RPP.
- Finally, the FS of the reinforced slope can be determined using Equation 5.12.
- If the FS of the reinforced slope is lower than the target FS, it is suggested that the spacing of the RPPs be reduced to increase the FS of the reinforced slope.

5.7.2   Approach 2: infinite slope

The FS of the surficial failure can also be deduced using the infinite slope method, with seepage parallel to the slope face. The schematic of infinite slope failure for c-φ soil with parallel seepage is presented in Figure 5.20.

If there is seepage through the soil and the water level coincides with the ground surface, the shear strength of the surface is given by:

$$\tau = c' + \sigma' \tan \varphi' \tag{5.13}$$

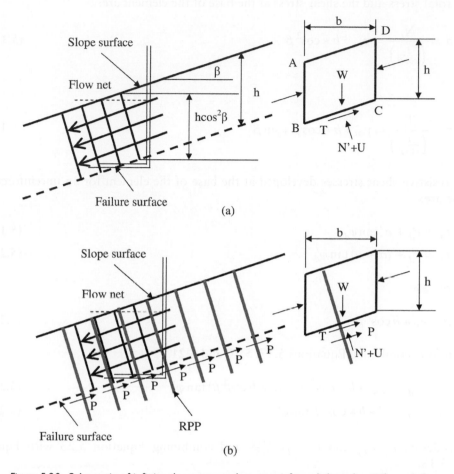

Figure 5.20 Schematic of infinite slope approach: a. unreinforced slope; b. reinforced slope.

Considering the slope elements A, B, C and D, the forces that act on the vertical faces of A and B and C and D are equal and opposite. The total weight of the slope element of unit length is:

$$W = \gamma_{sat} * L * h \tag{5.14}$$

where γ_{sat} is the saturated unit weight of soil.

The components of W in the directions of normal and parallel to plane AB are:

$$N_a = N_r = W\cos\beta = \gamma_{sat} * L * h * \cos\beta \tag{5.15}$$

and

$$T_a = T_r = W\cos\beta = \gamma_{sat} * L * h * \cos\beta \tag{5.16}$$

The total stress and the shear stress at the base of the element are:

$$\sigma = \frac{N_r}{\left(\frac{L}{\cos \beta}\right)} = \gamma_{sat} * h * \cos^2 \beta \tag{5.17}$$

and

$$\tau = \frac{T_r}{\left(\frac{L}{\cos \beta}\right)} = \gamma_{sat} * h * \cos \beta * \sin \beta \tag{5.18}$$

The resistive shear stresses developed at the base of the element for an unreinforced slope are:

$$\tau_d = c'_d + \sigma' \tan \varphi'_d \tag{5.19}$$
$$\tau_d = c'_d + (\sigma - u) \tan \varphi'_d \tag{5.20}$$

where

$$u = \gamma_w * h \cos^2 \beta \tag{5.21}$$

Therefore, combining Equations 5.17, 5.20 and 5.21:

$$\tau_d = c'_d + (\gamma_{sat} * h * \cos^2 \beta - \gamma_w * h \cos^2 \beta) \tan \varphi'_d \tag{5.22}$$
$$\tau_d = c'_d + \gamma' * h * \cos^2 \beta \tan \varphi'_d \tag{5.23}$$

Considering $c'_d = \frac{c'}{FS}$ and $\tan \varphi'_d = \frac{\tan \varphi}{FS}$ and combining Equation 5.23 with Equation 5.18:

$$FS = \frac{c' + h\gamma' \cos^2 \beta * \tan \varphi'}{\gamma_{sat} * h * \sin \beta * \cos \beta} \tag{5.24}$$

For reinforced slopes, as presented in Figure 5.20, RPPs provide additional resistance, P, along the base, and increase the shear resistance as:

$$\tau_d = c'_d + \gamma' * h * \cos^2 \beta \tan \varphi'_d + P/L \tag{5.25}$$

Therefore, for a reinforced slope, by combining Equation 5.25 with Equation 5.18, the FS can be determined as:

$$FS = \frac{c'L + hL\gamma' \cos^2 \beta * \tan \varphi' + \left(\frac{L}{s} + 1\right) * P}{\gamma_{sat} * hL * \sin \beta * \cos \beta} \tag{5.26}$$

where L = length (parallel to slope face), s = RPP spacing, and P = limit resistance of RPP.

5.7.2.1 Design steps for approach 2

The infinite slope is the simplest approach to determining the FS of both reinforced and reinforced slopes and can be performed using a simple Excel spreadsheet. The design can be performed with the following steps:

- The FS of the unreinforced slope should be determined using the soil parameters and depth of surficial failure. If no data is available on the depth of the surficial failure, it is suggested that the maximum failure depth of 7 ft reported by Loehr and Bowders (2007) be used.
- Using the depth of failure, the limit resistance, P, should be determined from the design chart, which is attached in Appendix A. The design chart should be utilised to determine the lowest resistance, based on the limit soil resistance, limit horizontal displacement and maximum flexure criteria, while considering the soil parameters and the structural parameters of the RPPs.
- The spacing of the RPPs should be considered.
- Finally, the FS of the reinforced slope can be determined by using Equation 5.26.

If the FS of the reinforced slope is lower than the targeted FS, it is suggested that the spacing of the RPPs be reduced to increase the FS of the reinforced slope.

Several sample calculations to use in designing slope stabilisation using RPPs are included in Appendix B.

5.8 DESIGN RECOMMENDATIONS

Several field studies are being conducted using RPPs for slope reinforcement. Based on the field performance results, the following design recommendations are made.

5.8.1 Extent of reinforcement zone

Field studies indicate that soil movement in the control slope may have lateral extent up to 25 ft. Therefore, it is recommended that the reinforced zone be extended at least 25 ft more from the edge of the failed area. The extent of the reinforcement zone is illustrated in Figure 5.21. If the slope experiences recurring failures or has different failure zones, it is suggested that the reinforcement zone be extended until the slope ratio reaches 1:4.

5.8.2 Material selection

RPPs are made from recycled plastics. Because of the wide variety in the sources, the composition of RPPs varies significantly. In addition, RPPs are polymeric materials, which are generally susceptible to creep. However, glass and wooden fibres are added as reinforcement during the manufacturing process of RPPs to reduce their vulnerability to creep failure. Therefore, it is highly recommended that commercially produced RPPs are selected as structural elements and reinforced with glass/wooden fibres. The minimum engineering properties required of RPPs are presented in Table 5.6.

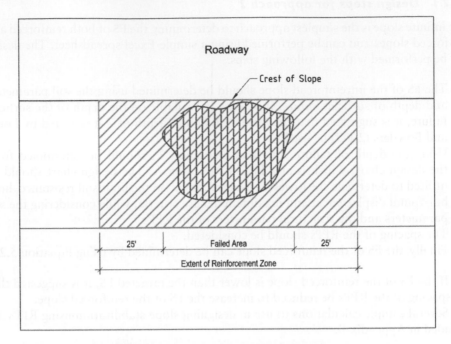

Figure 5.21 Extent of the reinforcement zone.

Table 5.6 Properties of RPPs.

Properties	ASTM test	Value (psi)
Flexure strength	D6109	2750
Flexural modulus (secant @ 1% strain)	D6109	288000
Compressive strength (parallel to grain)	D6108	2800
Compressive modulus (secant @ 1% strain)	D6108	54000
Shear strength	D2344	800

Table 5.7 Sectional requirements of RPPs.

Properties	Value
Minimum cross-sectional area	12.25 inch^2
Minimum moment of inertia along slope movement	12.5 inch^4

    The field study was conducted using a 4 inch by 4 inch square section. The section was effectively based on the installation and performance criteria. However, different rectangular and circular sections are commercially available, which may be used for slope stabilisation. Other sections of RPP can be utilised during the design; however, the minimum cross-sectional area and moment of inertia, as presented in Table 5.7, should be maintained.

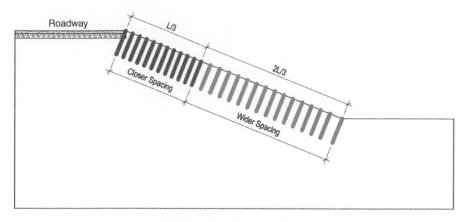

Figure 5.22 Layout of RPPs.

5.8.3 Selection of RPP spacing

Slope failures typically originate near the weakest zones: the crest of the slope and along the pavement shoulder. It was observed during the field study that reinforcement near the crest provided additional resistance against failure initiation and improved the performance of the slope significantly. Therefore, it is recommended that closer spacing is provided near the crest of the slope. This closer spacing of RPPs can be within the range of 2 to 4 ft, but 3 ft is the optimal and recommended spacing. The RPP spacing should be at least one-third of the slope length.

A wider spacing of RPPs can be provided in the remaining zones of the slope. The suggested spacing of RPPs ranges between 3 and 5 ft, with a spacing of 4 ft recommended at the bottom part of the slope. However, the maximum spacing of RPPs should not exceed 5 ft. The suggested layout of RPPs is illustrated in Figure 5.22.

5.8.4 Minimum RPP length and RPP sections

The design chart was developed on the basis of the length of the RPPs being 10 ft. However, RPPs longer than 10 ft can be used, but may result in a conservative design.

5.8.5 Recommendations on design method

Two different approaches to determining the FS of the control and reinforced slope were presented earlier, using the performance-based approach. The infinite slope method is widely used to determine the FS for shallow slope failures. Therefore, it is suggested that Design Approach 2 (infinite slope approach) be used to perform the slope remedial design. The following design criteria are recommended:

- Minimum FS of the reinforced slope = 1.5
- Maximum allowable horizontal displacement of RPPs = 2.0 in
- Depth of failure surface to be assumed if no site investigation result is available = 7 ft.

Figure 5.22 Layout of RPFs.

5.8.3 Selection of RPP spacing

Slope failure typically originate near the weakest zones: the crest of the slope and along the pavement shoulder. It was observed during the field study that reinforcement near the crest provided additional resistance against failure initiation and improved the performance of the slope significantly. Therefore, it is recommended that closer spacing is provided near the crest of the slope. This closer spacing of RPFs can be within the range of 2 to 4 ft, but 3 ft is the optimal and recommended spacing. The RPP spacing should be at least one-third of the slope length.

A wider spacing of RPFs can be provided in the remaining zones of the slope. The suggested spacing of RPFs ranges between 3 and 5 ft, with a spacing of 4 ft recommended at the bottom part of the slope. However, the maximum spacing of RPF should not exceed 5 ft. The suggested layout of RPFs is illustrated in Figure 5.22.

5.8.4 Minimum RPF length and RPF sections

The design chart was developed on the basis of the length of the RPFs being 10 ft. However, RPF longer than 10 ft can be used, but may result in a conservative design.

5.8.5 Recommendations on design method

Two different approaches to determining the FS of the control and reinforced slope were presented earlier using the performance-based approach. The infinite slope method is widely used to determine the FS for shallow slope failures. Therefore, it is suggested that Design Approach 2 (infinite slope approach) be used to perform the slope remedial design. The following slope design criteria are recommended:

- Minimum FS of the reinforced slope = 1.5
- Maximum allowable horizontal displacement of RPFs = 1.0 in
- Depth of failure surface to be assumed if no site investigation result is available = 7 ft.

Chapter 6

Construction methods

6.1 EARLY DEVELOPMENT OF CONSTRUCTION TECHNIQUES

A series of laboratory and field tests were performed using RPPs to evaluate different installation methods. Both impact methods and vibratory methods were considered. Sommers et al. (2000) evaluated the impact-driving technique in the laboratory, using a simple drop-weight driving mechanism to drive a small-scale 4 cm × 4 cm RPP into a soil-filled drum. The laboratory drive test indicated that RPPs are extremely resilient to driving stresses; however, this installation process was determined to be unsuitable for RPPs.

A further evaluation of installation technique was conducted using the vibratory driving method. In the course of this method, a slightly modified 27 kg (60 lb) pavement breaker was used to drive reduced-scale pins at a field test site near Columbia, Missouri (Sommers et al., 2000). The field test results indicated the resilience of RPPs and also demonstrated that the penetration rates for the pseudo-vibratory method far exceeded those of the drop hammer method.

Sommers et al. (2000) recommended the pseudo-vibratory method for subsequent full-scale RPP installations and conducted further field-scale tests, using the pseudo-vibratory mechanism at a site in St. Joseph, Missouri. The site was located in a flood plain, and had two soil layers. The upper layer was approximately 1.5 m (5 ft) of compacted low-plasticity clay. The next layer of soil was a natural alluvial deposit of highly plastic clay. The *in situ* dry densities of the site soils ranged from 86 to 107 lbs/ft^3. A total of seven 10 × 10 cm square pins of 1.2 m and 2.4 m lengths were driven at the site. A modified Indeco MES 351 hydraulic breaker mounted on a rubber-tyred Bobcat® 835 skid loader was used as the trial equipment, as shown in Figure 6.1. In the field trials, the penetration rates varied from a maximum of 12 ft/min to a minimum of 0.8 ft/min due to varying soil conditions. The penetration rate was highest in the soft, highly plastic alluvial deposits, with an average dry density of 86 lbs/ft^3. The lowest penetration rates were observed for highly compacted low-plasticity clays, with dry densities up to 106 lbs/ft^3. The penetration rate was considered effective for large-scale field installation, although the method had some limitations. The short wheelbase caused equipment stability problems during installation, and could prove unsafe for installation on slopes. In addition, the minimal headroom restricted the capacity to install RPPs of more than 2.4 m (7.9 ft) in length.

Figure 6.1 Evaluation of RPPs installation techniques at field scale (Sommers et al., 2000).

Based on the laboratory- and field-scale installation trials, it was evident that the pseudo-vibratory method provided some advantages. However, due to the limitations of the tyre-mounted equipment, crawler-mounted systems were selected for further trials and proved to be an appropriate method for installation. Descriptions of the equipment that was successfully used for field installations are presented in the following section.

6.2 TYPES OF EQUIPMENT AND DRIVING TOOLS FOR FIELD INSTALLATION

Several field installations have been conducted using both the pseudo-vibratory method and the impact-driving technique. The field installations at the test sites indicated that RPPs can be installed using either the driving or vibratory method, at a reasonable driving rate. This section describes some of the most appropriate equipment and its use in slope stabilisation.

6.2.1 Davey Kent DK 100B drilling rig

The installation of RPPs was performed using this equipment at a field demonstration site in Kansas City, Kansas, during November 1999. The rig was mounted on a crawler and equipped with a mast capable of a 50-degree tilt forward from vertical, 105-degree tilt backward, and side-to-side tilt of 32 degrees from vertical. The installation technique indicated that the mast system was very effective at maintaining the alignment of the hammer and the crawler along the same line. This significantly reduced the chance that eccentric force would develop, and thereby reduced any buckling of the RPPs.

Figure 6.2 DK 100B crawler-mounted drilling rig and mast system used for installation of RPPs (Sommers et al., 2000).

The crawler-mounted rig also had another significant benefit, which was that it provided stability during the driving and was much easier to manoeuvre on the inclines (Sommers et al., 2000). As a result, the time required to move the equipment from one point to another was significantly less. The rig was equipped with a Krupp HB28A hydraulic hammer drill attached to the mast. The hydraulic hammer was capable of producing a maximum of 400 N-m (295 ft/lbs) of energy at a maximum frequency of 1,800 blows/min. The hammer energy was further enhanced by a push/pull of 8,165 kg (18,000 lbs) supplied by the drill mast. A field installation photo, taken while using this drilling rig, is presented in Figure 6.2. The field installation technique indicated that RPPs installed in a vertical alignment were driven with the rig being backed up the slope. As a result, this feature lowered the installation rates for pins driven vertically, as compared to pins driven perpendicular to the face of the slope.

6.2.2 Klemm 802 drill rig with KD 1011 percussion head drifter

The installation of RPPs was performed using this rig at several field demonstration site locations around Dallas, Texas, during March 2011. This specific type of rig was selected according to the successful outcome from the DK 100B drilling rig. The Klemm 802 is a compact and multi-purpose drilling rig. The standard boom with its $6 \times 90°$ swivel head allows the highest possible flexibility. It is equipped with an 18 ft long mast, which allows installation of RPPs of length up to 15 feet. The KD 1011 drifter is a hydraulic hammer which is equipped with two motors and can produce a blow frequency of 2,800 blows/min. The single-blow energy for each of the KD 1011 drifters is 295 ft/lb.

Figure 6.3 Klemm 802 drill rig with a KD 1011 percussion head drifter.

The crawler-type rig is suitable for installation over slopes, as no additional anchorage is required to maintain the stability of the equipment. This reduces the amount of labour, cost and time required for the installation process. Photographs of RPP installation using the Klemm 802 drill rig along with a KD 1011 percussion drifter are presented in Figure 6.3.

Using the Klemm 802 drill rig, the installation of the RPPs typically started from the crest of the slope. During the field demonstration process, it was observed that the crawler made it easy to manoeuvre the equipment on the slope, and the set-up time was significantly less when the installation started from the crest and the rig was gradually backed down towards the toe of the slope. In addition, during this process, the depression of the ground due to the crawler movement is much smaller.

6.2.3 Deere 200D with FRD F22 hydraulic hammer

This crawler-mounted rig with mast equipped with a pseudo-vibratory hammer works well for RPP installation. However, one major limitation is that this rig is not widely available and requires a special operator to install the RPPs. A crawler-mounted rig with a pseudo-vibratory hammer (model: Casagrande M9-1) was used to install the RPPs in a slope stabilisation project in Texas. However, the rig was not suitable due to the steepness of the slope at the crest. At the highest steepness (2.5H:1V) near the crest of the slope, the crawler of the rig tended to tilt and lose ground contact during the RPP-driving process. As a result, this rig was replaced with a crawler-mounted excavator which had greater stability on steep ground.

These crawler-mounted excavators, with extendable booms, offered several benefits. They had additional reach, which allowed the equipment to remain off the slope during RPP installation, further limiting damage to the slope and reducing the set-up time. Compared to a track-mounted system, an excavator with an extendable boom also has a greater swing range, which allowed a larger number of RPP installations without movement of equipment, and reduced set-up time.

Figure 6.4 RPP installation using Deere 200D with an FRD F22 hydraulic hammer.

The Deere 200D is a medium-size excavator with a net power of 159 hp. The excavator was equipped with a hydraulic system, which could facilitate a hydraulic flow of 112 gal/min, and a Furakawa Rock Drill (FRD) F22 hydraulic hammer capable of producing an impact energy of 4,000 lbs/ft, with between 360 and 700 blows per minute. The F22 hydraulic hammer produces an impact energy 13.5 times higher than the KD 1011 percussion drifter, and requires a minimum hydraulic flow of 37 gal/min. The hammer needs a minimum pressure of 2,320 psi to operate, which requires the RPPs to be pushed into the ground to trigger the hammer impact for installation. During the field installations, stiff soil was encountered at some locations, and pushing the RPPs into the ground using the powerful hammer caused buckling and cracking of the RPPs.

This method was useful on soft ground, as the excavator was capable of pushing most of the RPPs. However, another limitation of the equipment is that the excavator does not have a mast system. As a result, it relied on the expertise of the rig operator to maintain vertical alignment. At the beginning of installation, the wastage of RPPs was high; however, with time, as the operator became used to the installation technique, the wastage decreased significantly. Overall, a total of 130 RPPs were installed with a wastage of less than 5%. Photographs of this RPP installation are presented in Figure 6.4.

6.2.4 Caterpillar CAT 320D LRR with CAT H130S hydraulic hammer

More than 500 RPPs were installed in two different slopes in Dallas, Texas, using a CAT 320D LRR excavator, which is a medium-sized excavator with a net power of 148 hp, and features similar to the Deere 200D. The excavator was equipped with a hydraulic system that can produce a hydraulic flow of 54 gal/min and can generate a maximum lifting pressure up to 5,221 psi. For the RPP installation, the excavator

Figure 6.5 RPP installation using CAT 320D LRR with a CAT H130S hydraulic hammer.

was equipped with a CAT H130S hydraulic hammer capable of producing an impact energy of 4,500 lbs/ft, with between 320 and 600 blows per minute.

A photograph of the RPP installation is presented in Figure 6.5. Similarly to the F22 hydraulic hammer, this CAT hammer requires a minimum pressure (2,030 psi) to operate, which requires the RPPs to be pushed into the ground to trigger the hammer to begin the installation. The impact-driving method, used in this system, was valuable in the soft ground as the excavator was capable of pushing in most of the RPPs. However, like the Deere 200D, this equipment doesn't have a mast system, which made it difficult to drive the pins cleanly at the beginning, although as the rig operator became used to maintaining the alignment, the installation rate increased significantly.

A slope stabilisation project using a CAT 320D was undertaken during the dry summer, and it was very difficult to drive the RPPs into the hard soil layer. To overcome this problem, a full-sized steel pin was manufactured and driven into the predefined locations. The steel pin was then withdrawn, leaving an empty hole and making it easy to push in the RPP. This technique proved to be very effective in the dry clay soil and is discussed further at the end of this chapter.

6.3 FIELD INSTALLATION RATE

During the installation of RPPs, the installation time was recorded to investigate the penetration and installation rates. The penetration rate for the RPPs was considered as the time taken to drive the RPPs a given distance into the ground, and the installation

Table 6.1 Summary of penetration and installation rates from an RPP installation project in Missouri (Sommers et al., 2000).

Rate	Penetration rate (ft/min)	Installation rate (ft/min)
Average	4.64	1.37
Maximum	10	2.1
Minimum	0.12	0.55

rate was considered as the total combined time required to drive the RPPs and to set up the equipment before installation. The crawler-mounted rig was easy to operate in the inclined ground and required no additional anchorage, thereby significantly reducing the set-up time and increasing the installation rate of the RPPs.

Sommers et al. (2000) reported that the maximum penetration rate for driving RPPs perpendicular to the slope was 10 ft/min, with an average rate of 5.2 ft/min. Penetration rates for RPPs driven vertically were only slightly lower, reaching 9.6 ft/min and averaging 4.1 ft/min. It was also observed that it was easier to install the RPPs perpendicular to the ground surface, as it provided more stability for the rig. Installation rates were also faster, with a maximum of 2.1 ft/min at peak production. The average installation rate for the installation of all pins was 1.33 ft/min. The expertise of the rig operator played a major role in determining the installation rates, which generally increased with time as the construction team became familiar with the installation process. The installation rates for the field demonstration study performed in Missouri are summarised in Table 6.1.

Several RPP installation projects from 2012–16 in the Dallas–Fort Worth area of Texas used both the vibratory method and the impact method. A crawler-mounted rig equipped with a mast-mounted pseudo-vibratory hammer (Klemm 802 drill rig with KD 1011 percussion head drifter) was used to install more than 600 RPPs in a highway slope stabilisation project. With a 3 ft centre-to-centre (c/c) spacing for RPPs 10 ft in length, the RPP installation rate was 2.85 ft/min. The installation rate reduced to 2 ft/min when RPP spacing was increased to 6 ft c/c, due to the longer time required to manoeuvre the equipment between the higher spacings of the RPPs. Conversely, the highest driving rate of 3 ft/min was observed with 8 ft long RPPs at 5 ft c/c spacing, located near the toe of the slope. The soil near the toe of the slope was very soft and, as a result, the time it took to drive the RPPs into the slope reduced dramatically, producing the highest driving rate. The overall average installation rate for the section was observed as 2.72 ft/min.

The installation of 10 ft long RPPs at 4 ft c/c spacing was conducted at a different section, near the end of the three-day installation process. The construction team had become experts at installing the RPPs, which significantly reduced the time required for installation, as depicted in Table 6.2.

Based on the pilot study in Texas, the average driving rate was 2.66 ft/min, which signifies that a 10 ft long RPP can be installed within four minutes, including the manoeuvring of the crawler-mounted equipment. At this installation rate, on average, a total of 100–120 RPPs could be installed per day.

Table 6.2 Average RPP installation rate in a slope stabilisation project in the Dallas–Fort Worth area of Texas.

Length of RPP (ft)	RPP spacing (ft)	Average RPP-driving time (min)	Average RPP-driving rate (ft/min)
10	3	3.55	2.9
10	4	2.76	3.6
10	6	4.76	2.1
8	4	3.08	2.6
8	5	2.63	3.1
8	6	3.65	2.2

6.4   POTENTIAL CHALLENGES OF RPP INSTALLATION

Several field installations have already been conducted in Missouri, Iowa and Texas. Some of the challenges observed during the RPP installation process are discussed in the following subsections.

6.4.1   Slope steepness

The steepness of the slope has an influence on the RPP installation process because it can cause instability of the rig. During the installations in North Texas, it was observed that the crawler-mounted rig with a mast system was not suitable for installations on steep slopes because the vibration of the equipment caused it to lose contact with the ground. Moreover, the movement of the mast caused additional movement, during which the rig tended to tilt. This condition worsened during bad weather. Due to rainfall, the slope, which was constructed using highly plastic clay soil, became saturated and soft, creating unfavourable conditions for operating equipment on its surface. Slopes on soft soil became slippery, the crawler system was unable to grip the ground and it became unstable. A crawler-mounted excavator with an extendable boom was more stable on steep slopes.

6.4.2   Skilled labour

Slope stabilisation using RPPs is a sustainable technique. It is important to note that RPPs are made of plastic and have a lower elasticity modulus than other alternative construction materials, such as steel. As a result, installation of RPPs using powerful driving equipment is not straightforward. Some trials have been run in which RPPs have been driven into the ground using a heavy jackhammer mounted on an excavator. In this case, it was observed that the installation of the RPPs required a gentle push and drive into the ground, rather than being hammered at the top. Repeated hammering on the top of an RPP can often cause buckling and permanent deformation. A new construction team may also lack experience in manoeuvring a crawler-mounted excavator, resulting in lower RPP installation rates (<0.75 ft/min) and high RPP wastage (20–25%). Coordination among members of the construction team is also crucial and

Figure 6.6 Solid drive head for RPP installation.

influences the installation rate. As the construction team becomes familiar with the installation process, the installation rate increases noticeably (>3 ft/min), and wastage of RPPs can be reduced to less than 3%. As the use of slope stabilisation techniques using RPPs increases, more contractors will become adept at implementing the installation process. Until then, it is recommended that RPP-experienced contractors be employed for all similar projects, where possible.

6.4.3 Connection between the hammer and pile head

Several mechanical problems slowed installation progress at the demonstration site in Texas. For example, a drive head was fabricated that served as a connection between the hammer and the RPPs. It failed at various points during the installation, thus slowing down the process.

In the early stages of installation with mast-mounted hammers, a welded drive head was used as a connector and allowed the transfer of energy from the hammer to the RPPs. Eventually, though, the repetitive impact from the hammer caused the welded connections to fail. Several spare connectors were kept on site but construction was slowed down due to repeated failure of the welded joints. Subsequently, a new hammer head was fabricated, using a solid stainless steel mould, without any joints or welded parts; it is pictured in Figure 6.6. The solid drive head worked well, and more than 300 RPPs were installed using it, without any interruption or sign of disturbance.

6.5 SPECIAL INSTALLATION TECHNIQUES IN ADVERSE SITUATIONS

Highly plastic clay soil absorbs and retains water during wet periods, which results in the top soil becoming soft. The vibratory and impact methods work very well

Buckling of RPP

Figure 6.7 Buckling and permanent deformation of RPP in dry soil.

Driving a steel pin into the ground

Driving an RPP into the hole

Figure 6.8 RPP installation at the SH 183 slope.

when installation is performed during a wet period and the top soil layer is relatively soft. However, during elongated summers, especially hot and dry ones like those in Texas, the clay soil loses moisture from the top, making the soil very stiff and causing unfavourable conditions for driving the RPPs into the ground.

A highway slope stabilisation programme was undertaken in North Texas during the month of September, immediately after an elongated summer where temperatures were greater than 90°F for more than two months. During the installation project, a Caterpillar CAT 320D LRR rig with a CAT H130S hydraulic hammer was used to drive RPPs into the ground. Because the soil was stiff during the dry weather, the installation rate reduced significantly (<0.5 ft/min), and buckling, permanent deformation and breakage of the RPPs increased significantly, as presented in Figure 6.7.

To overcome this problem, a full-sized steel pin was manufactured to drive holes into the predefined locations. After driving in the steel pin, it should be withdrawn immediately, leaving an empty hole into which an RPP can easily be pushed. A 7 ft long steel pin was welded to the drive head and attached to the hammer using a steel chain. The steel pin was driven into the intended RPP locations, to make a hole up to 7 ft deep in the stiff soil layer. The steel pin was then pulled out of the ground, and the RPP was pushed in immediately. This installation procedure worked well, and more than 500 RPPs were installed at an installation rate of more than 2 ft/min. Photographs of the RPP installation are presented in Figure 6.8.

Chapter 7

Case studies

Recycled plastic pins (RPP) were first used for slope stabilisation in Missouri and Kansas in 1999, and after that they were successfully used in many other locations within those states. RPPs were utilised for slope stabilisation in North Texas for the first time in 2011, by a research team from the University of Texas at Arlington (UTA). The UTA team worked with the Texas Department of Transportation (TxDOT), Dallas District to implement slope stabilisation using RPPs. Since then, four slopes in North Texas have been stabilised using RPPs. Some case studies of highway slopes in Texas and Missouri repaired using RPPs are presented in this chapter.

7.1 US HIGHWAY 287 SLOPE IN MIDLOTHIAN, TEXAS

This fill slope, constructed during 2003–2004, is located at US Highway 287, near the St. Paul overpass in Midlothian, Texas. A photo of the slope appears in Figure 7.1. The maximum slope height is about 30–35 ft, with a slope geometry of 3H:1V. During September 2010, cracks were observed on the shoulder, near the crest of the highway slope.

Figure 7.1 Site photo of the US 287 slope, Texas (Khan et al., 2016).

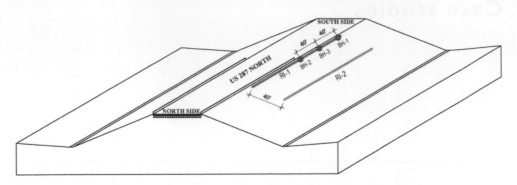

Figure 7.2 Layout of boreholes and resistivity imaging lines for US 287.

7.1.1 Site investigation

A subsurface exploration programme was conducted in October 2010 to obtain information about the site's soil and its possible path of movement. Two two-dimensional (2D) resistivity imaging (RI) tests were conducted at the site, and the layout of the boreholes and RI lines are presented in Figure 7.2.

A total of three soil test borings were performed near the crest of the slope, at depths ranging from 20 to 25 ft. Both disturbed and undisturbed soil samples were collected from different depths and tested to determine the geotechnical properties of the subsoil. Based on the laboratory investigation results, all of the collected soil samples were classified as high-plasticity clay (CH) soil, according to the Unified Soil Classification System (USCS). The liquid limits and the plasticity indices of the samples ranged between 48–79 and 25–51, respectively. The moisture profiles by depth and the plasticities along the three boreholes are presented in Figure 7.3.

The project site was located in the Eagle Ford geological formation, which is composed of residual soils consisting of clay and weathered shale (shaly clay), underlaid by unweathered shale. The weathered shale contains gypsum infills and debris, jointed and fractured with iron pyrites. The unweathered shale is typically grey to dark grey and commonly includes shell debris, silty fine sand particles, bentonite and pyrite. The Eagle Ford formation consists of sedimentary rock that is in the process of degrading into a soil mass. This formation also contains smectite clay minerals and sulfates. It should be noted that the smectite clay minerals are highly expansive in nature.

Two-dimensional RI is extensively used in shallow geophysical investigations and geo-hazard studies (Hossain et al., 2010). RI provides a continuous profile of the subsurface and the continuity of soil and moisture profiles throughout the site. During this study, the RI test was used to investigate the subsurface condition of the US 287 slope. The RI investigations were conducted using eight-channel AGI SuperSting equipment. A total of 56 electrodes were utilised during the RI. The length of the investigated line was 275 ft, with electrode spacing of 5 ft centre-to-centre.

The 2D RI profiles along lines RI-1 and RI-2 (see Figure 7.2) are presented in Figures 7.4a and 7.4b, respectively. Based on the profile, a low-resistivity zone was observed near the top soil at both RI-1 (at crest) and RI-2 (middle of the slope). The resistivity of soil on the slope was less than 3 ohm at a depth of between 5 and 12 ft.

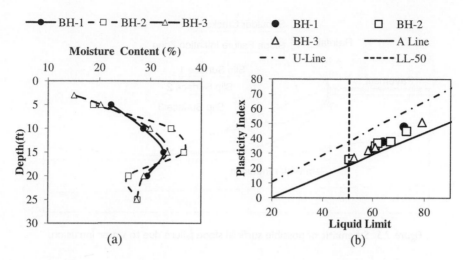

Figure 7.3 Laboratory test results: a. moisture distribution along the boreholes; b. plasticity along the boreholes.

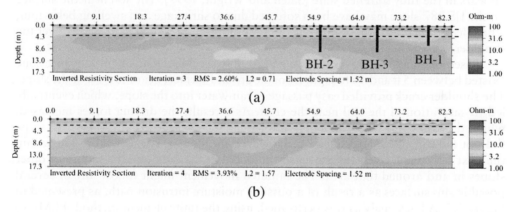

Figure 7.4 Resistivity imaging at the US 287 slope: a. resistivity profile for RI-1; b. resistivity profile for RI-2.

It should be noted that low resistivity indicates the presence of high moisture levels in the soil; usually, the presence of high moisture levels in soil increases electrical conductivity and reduces electrical resistance. The moisture profile matched the site investigation results, which used soil boring logs.

The subsoil investigation results indicated that the US 287 slope was constructed using high-plasticity clay. In addition, the dominant mineral of the soil is montmorillonite. In the presence of montmorillonite, high-plasticity clay is highly susceptible to swelling and shrinking upon wetting and drying. It should be noted that fully softened strengths are eventually developed in high-plasticity clays in the field after exposure to environmental conditions (i.e. shrink and swell, wetting-drying, etc.), and they represent the governing strength for first-time slides in both excavated and fill slopes (Saleh and Wright, 1997). The reduction in the friction angle is not significantly affected, due

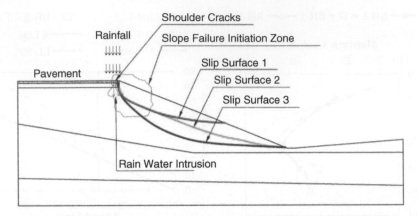

Figure 7.5 Schematic of possible surficial slope failure due to water intrusion.

to cyclic wetting and drying of the soil; however, the cohesion of the soil almost dis-appears in the fully softened state (Saleh and Wright, 1997). The soil near the surface on the US 287 slope may have been softened due to shrinkage and swelling behaviour, which led to the initiation of movement of the slope and resulted in the crack over the shoulder.

Based on the subsoil investigation and RI, it was evident that a high-moisture zone existed between 5 ft and 14 ft depth from the ground surface, near the crest of the slope. The shoulder crack provided easy passage of rainwater into the slope, which eventually led to saturation of the soil near the crest. As a result, the driving forces increased, which decreased the factor of safety (FS). It should be noted that the US 287 slope did not fail during the investigation. However, the slope was expected to fail within the next few years, given the movement initiated at the crest and the failure of other slopes in and around that area. The initiated movement might follow any of several possible slip surfaces as a result of a possible moisture intrusion path, as presented in Figure 7.5. A back analysis was performed, using the finite element method (FEM), to evaluate the critical shear strength at an FS equal to 1.

7.1.2 Slope stability analyses at US 287 slope

Slope stability analysis by the elasto-plastic FEM is accurate, robust and simple. In addition, the graphical presentation of an FEM program allows better understanding of the slope failure mechanism. The slope stability analyses of the US 287 slope were performed using the FEM program, PLAXIS 2D (Plaxis, 2011). The Mohr–Coulomb soil model was utilised for stability analysis, using 15-node triangular elements. The 15-node triangle is a very accurate element and produces high-quality stress results for different problems. Standard fixities were applied as a boundary condition, where the two vertical boundaries were free to move vertically and were considered fixed in the horizontal direction. The bottom boundary was modelled as fixed.

In the FEM back analysis, it was assumed that the initiation of slope movement would take place with a limiting FS equal to 1.0. The shear strength reduction method

Table 7.1 Soil parameters at slope failure.

Soil type	Friction angle φ (°)	Cohesion c (psf)	Unit weight γ (pcf)	Elastic modulus E (psf)	Poisson ratio ν
1	10	100	125	100000	0.35
2	23	100	125	150000	0.3
3	15	250	130	200000	0.25
4	35	3000	140	250000	0.2

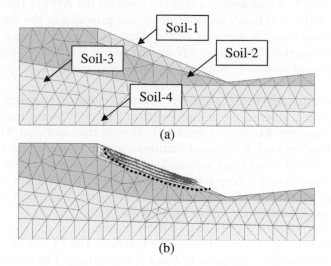

(a)

(b)

Figure 7.6 Slope stability analysis using PLAXIS 2D software: a. soil model; b. critical slip surface for factor of safety = 1.05.

was used to determine the FS. The FS of a soil slope is defined in the shear strength reduction method as the factor by which the original shear strength parameters can be reduced in order to bring the slope to the point of failure.

The soil profile for the model is presented in Figure 7.6a. The top seven feet of soil was considered as a failure zone, with fully softened strength. Other soil parameters for different soil layers were taken from field investigation results. Several iterations were performed in the course of the numerical analysis to derive the soil parameters at slope failure, and these parameters are presented in Table 7.1. Based on the FEM analysis, the FS of the slope was found to be 1.05, as illustrated in Figure 7.6b.

The FEM analysis indicated a typical pattern of shallow slope failure, which resembled the displacement trend observed at the US 287 slope. The lowest FS along the failure plane was estimated as 1.05.

7.1.3 Slope stabilisation using RPPs

Three sections over the US 287 slope, designated as Reinforced Sections 1, 2 and 3, were considered for stabilisation. In addition, two unreinforced control sections that were between the reinforced sections were considered in order to evaluate the performance of the latter. The width of each section was 50 ft.

During rainfall, water entered the slope through the crack over the shoulder and saturated the soil near the crest. As a result, the pore water pressure in the saturated zone increased, decreasing the shear strength. It was evident that the cracked zone was the initiation point of the critical slip surface of the US 287 slope. Therefore, to resist the movement of the slope and provide additional support, RPPs with closer spacing were installed at the crest of the slope in Reinforced Section 1. Thus, 10 ft long RPPs, at a centre-to-centre (c/c) spacing of 3 ft, were proposed near the crest of the slope in Reinforced Section 1. It was assumed that the force on the RPPs in the middle of the slope might be reduced if heavy reinforcement was provided in the top section, and that the movement from the top part of the slope could be minimised by installing the RPPs closer together. Therefore, 6 ft c/c spacing of RPPs was proposed for the middle of Reinforced Section 1. Near its toe, a 5 ft c/c spacing was proposed, with 8 ft long RPPs.

Different lengths of RPPs (10 ft at the crest and 8 ft near the toe) were proposed for Reinforced Section 2, whereas a constant length of RPP (10 ft) was proposed throughout the entire slope for Reinforced Section 3. During this study, an RPP spacing of 4 ft c/c was utilised for both Reinforced Sections 2 and 3. However, the first two rows of RPPs at the crest of Reinforced Section 3 were proposed with 3 ft c/c spacing, and were installed more than 2 ft deep for the existing slope surface. RPPs were placed in a staggered grid over the reinforced sections. The proposed layout and cross section of each of the sections are presented in Figure 7.7.

Based on the proposed distribution of RPPs, further slope stability analyses were conducted to evaluate the FS of each reinforced section. The FS was observed as 1.43, 1.48 and 1.54 for Reinforced Sections 1, 2 and 3, respectively. The critical slip surfaces for each of the reinforced sections are presented in Figure 7.8.

7.1.4 Field installation

Sommers et al. (2000) performed a study of different construction techniques that could be used to install RPPs in the field. They concluded that the mast-mounted pseudo-vibratory hammer system works well for field installation because it maintains the alignment of the hammer and restricts the imposition of additional lateral loads during the RPP driving process (Bowders et al., 2003). Therefore, a similar crawler-type drilling rig with a mast-mounted vibratory hammer (model: Klemm 802 drill rig with KD 1011 percussion head drifter) was utilised during the current study to install the RPPs. The crawler-type rig is suitable for the installation process over the slopes because no additional anchorage is required to maintain the stability of the equipment, which reduces the labour, cost and time required for the installation process. Photographs of the RPP installations at the US 287 slope are presented in Figure 7.9. The RPPs were installed in Reinforced Sections 1 and 2 in April 2011.

The RPP driving time was measured during the installation process. Based on this measurement, the average installation times, as well as the driving rates, are summarised in Table 7.2.

It should be noted that the installation time per RPP is the summation of the time required to install the RPP and to manoeuvre the rig to the next location. At Reinforced

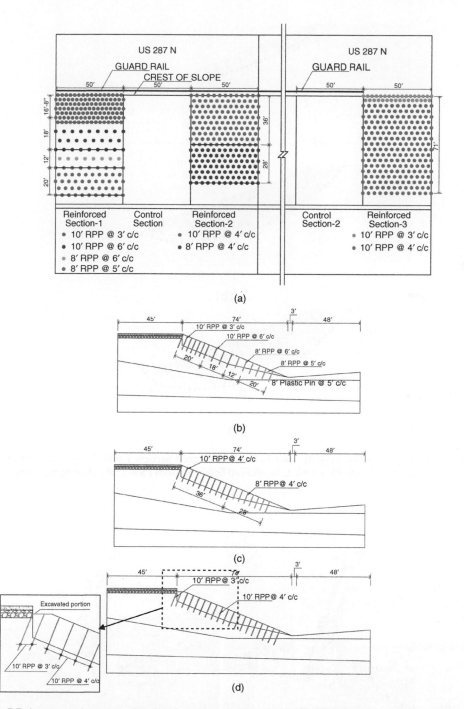

Figure 7.7 Layout and section details of slope stabilisation on US 287 slope: a. layout of RPPs; b. Reinforced Section 1; c. Reinforced Section 2; d. Reinforced Section 3.

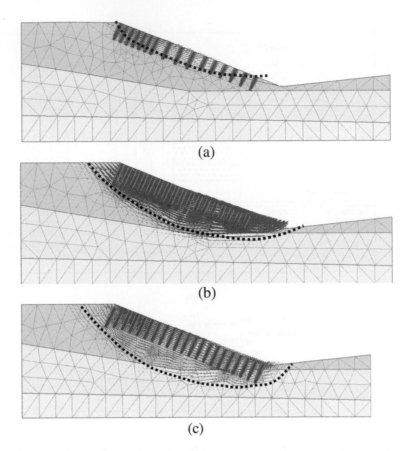

(a)

(b)

(c)

Figure 7.8 Slope stability analyses using RPPs: a. Reinforced Section 1 with FS = 1.43; b. Reinforced Section 2 with FS = 1.48; c. Reinforced Section 3 with FS = 1.54 (Khan et al., 2016).

Figure 7.9 Installation of RPPs in Reinforced Section 1 and Reinforced Section 2 (Khan et al., 2016).

Table 7.2 Average RPP driving time at US 287 slope.

Location of RPP	Length of RPP (ft)	RPP spacing (ft)	Average RPP driving time (min)	Average RPP driving rate (ft/min)
Reinforced Section 1	10	3	3.55	2.9
	10	6	4.76	2.1
	8	6	3.65	2.2
	8	5	2.63	3.1
Reinforced Section 2	10	4	2.76	3.6
	8	4	3.08	2.6
Reinforced Section 3	10	4	4.65	2.1

Section 1, the driving rate was observed as 2.85 ft/min at a 3 ft c/c spacing for 10 ft long RPPs. The driving rate was reduced to 2 ft/min along the middle of the slope, as RPP spacing increased to 6 ft c/c. The reduction in driving rate was due to the longer time required to manoeuvre the equipment between the wider spacings of the RPPs. Conversely, the highest driving rate of 3 ft/min was observed near the toe of Reinforced Section 1. During the installation process, the soil near the toe of Reinforced Section 1 was very soft. As a result, the installation time was reduced drastically, resulting in the highest driving rate. The overall average driving rate for Reinforced Section 1 was observed as 2.72 ft/min.

The driving rate was observed as 3.6 ft/min at the top of Reinforced Section 2, where the RPP spacing was 4 ft c/c. The driving rate was higher than that of Reinforced Section 1, as the installation team became more proficient. However, the driving rate was only 2.6 ft/min near the toe of Reinforced Section 2, due to the existence of a stiff foundation layer at that location. The overall driving rate for Reinforced Section 2 was observed as 3.18 ft /min.

The installations in Reinforced Section 3 were conducted in the following year (2012) and the RPPs were spaced similarly to those in Reinforced Section 2. However, a lower driving rate, of 2.13 ft/min, was observed. It should be noted that a new team installed the RPPs in Reinforced Section 3. Moreover, during installation, the new team experienced a stiff foundation soil beyond a depth of 7 ft, which resulted in a higher installation time. The installation process was also delayed due to mechanical problems with the driving equipment. As a result, the overall driving rate was low for Reinforced Section 3.

Based on the study, the average driving rate, considering all three reinforced sections, was 2.66 ft/min, indicating that a 10 ft long RPP can be installed within 4 minutes. Therefore, on average, a total of 100–120 RPPs can be installed in one day.

7.1.5 Instrumentation and performance monitoring

To evaluate the performance of the reinforced slope, selected RPPs were instrumented with strain gauges. Electrical resistance gauges of 350 ohm, with gauge lengths of both 0.5 in and 0.25 in, were utilised to instrument the RPPs. The strain gauges were fabricated with annealed constantan foil, along with tough, high-elongation polyimide

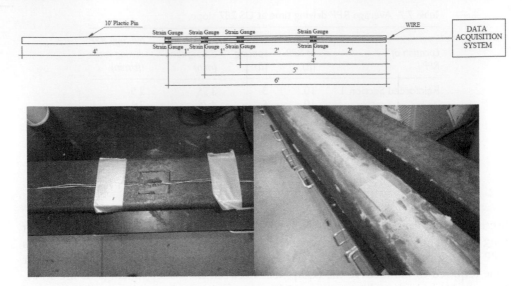

Figure 7.10 Installation of strain gauges on RPP.

backing, and were capable of measuring strain up to 20%. The gauges were placed on the RPPs in a recessed area, which was 0.25 in deep, to prevent them from being ripped off during installation. Gauges were attached using a special adhesive that was compatible with the RPP and cured almost instantly to produce an essentially creep-free, fatigue-resistant bond, with an elongation capability of 5% or more. The gauges were then sealed with waterproof sealant, and the recessed areas were filled using silicone caulk. The schematic and photograph of the installed strain gauges are presented in Figure 7.10.

A total of nine RPPs, designated as IM-1 to IM-9, were instrumented with strain gauges and installed in Reinforced Sections 1 and 2, and a Control Section. The instrumented RPPs were driven into the site during the field installation. The layout of the instrumentation at the US 287 slope is shown in Figure 7.11.

Three inclinometers, designated as Inclinometers 1, 2 and 3, were installed at Reinforced Section 1, Control Section and Reinforced Section 2, respectively, to monitor the horizontal movement of the slope. The depth of each inclinometer casing was 30 ft, and they were installed perpendicular to the slope surface, 20 ft below the crest. The locations of the inclinometers are presented in Figure 7.11.

A topographic survey was conducted over the US 287 slope as part of the performance monitoring of slope stabilisation. The first survey of the slope was conducted during May 2012, following the completion of RPP installation in Reinforced Section 3, and continued on a monthly basis. During the survey, the cracked zones over the shoulder, at different reinforced and control sections, were monitored. The layout of the survey lines is presented in Figure 7.12.

Instrumentation of the slope, using moisture sensors, water potential probes and a rain gauge, was carried out in the months of November and December 2010. Moisture sensors and water potential probes were installed at the crest and middle of the slope,

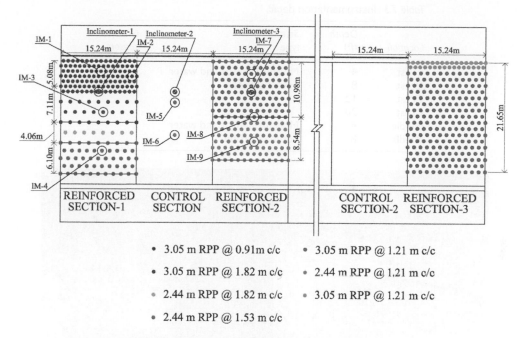

- 3.05 m RPP @ 0.91m c/c
- 3.05 m RPP @ 1.82 m c/c
- 2.44 m RPP @ 1.82 m c/c
- 2.44 m RPP @ 1.53 m c/c

- 3.05 m RPP @ 1.21 m c/c
- 2.44 m RPP @ 1.21 m c/c
- 3.05 m RPP @ 1.21 m c/c

Figure 7.11 Layout of instrumentation at US 287 slope.

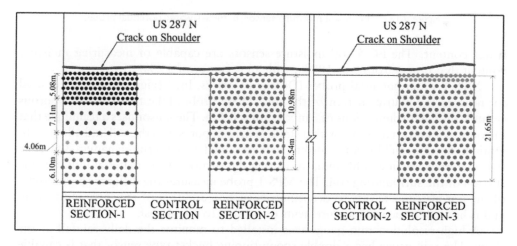

Figure 7.12 Layout of survey lines at US 287 slope.

at different depths, as summarised in Table 7.3. The sensors were connected to data loggers in the field to obtain continuous readings of the *in situ* moisture and suction.

Commercially available EC-5 soil moisture sensors (Decagon Devices, Inc.) (Figure 7.13a) were used to measure the volumetric water content of the soil. The overall dimensions of the probes were 3.5 in × 0.7 in × 0.3 in. The sensors measured the dielectric constant of the surrounding medium in order to establish the volumetric

Table 7.3 Instrumentation detail.

Location	Depth (ft)	Sensor type	No. of sensors
Crest	4	Moisture sensor and water	4
Crest	8	potential probe	
Crest	12		
Crest	20		
Middle	4	Moisture sensor and water	2
Middle	8	potential probe	

(a)   (b)

Figure 7.13 a. EC-5 soil moisture sensor; b. MPS-1 water potential probe.

water content. The EC-5 soil moisture sensors are capable of measuring moisture content that ranges from 0 to 100%, with an accuracy of ±2%.

MPS-1 water potential probes (Decagon Devices, Inc.) (Figure 7.13b) were used to measure the matric suction of the soil. These probes take the form of a porous ceramic disc, with dimensions of 3 in × 1.3 in × 0.6 in. The sensor uses a technique that introduces a known material with a static matrix of pores into the soil and allows it to attain hydraulic equilibrium according to the second law of thermodynamics. Once the two materials are in equilibrium, measuring the suction of the material indicates the suction of the surrounding soil. The MPS-1 probe measures the dielectric permittivity of the porous ceramic disc to determine its suction and that of the surrounding soil, and is limited to suction measurements of −208.9 to 10,445 psf.

A high-resolution rain gauge was installed to monitor the daily rainfall at the slope. The rain gauge has a double spoon-tipping bucket-type sensor that is capable of measuring rainfall amounts up to 0.08 in. The rain gauge was connected to a data logger placed in the field to record the amount of hourly rainfall; total daily rainfall is the summation of all the data recorded in a 24-hour period.

After the installation of sensors, the wires from the sensors were connected to an automatic data acquisition system to monitor the moisture content and matric suction on a continuous basis. Three Decagon Em50 data loggers were set up in the field to accommodate all of the sensors. The Em50 is a five-port, self-contained data logger that can measure data in continuous intervals. The measurement interval for the current

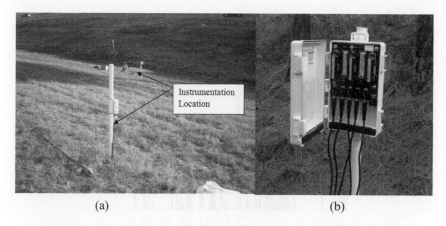

(a)   (b)

Figure 7.14 Data collection: a. instrumentation locations; b. Em50 data logger.

study was set to 60 minutes, which allowed data to be stored 24 times per day. The instrumented site, along with the data logger, is illustrated in Figure 7.14.

7.1.6 Performance of RPPs based on the results from instrumentation

Instrumented RPPs, IM3, IM5 and IM8, were installed in the middle of the slope in Reinforced Section 1, Control Section and Reinforced Section 2, respectively. Results obtained from these instrumented RPPs are presented in Figure 7.15. Based on the monitoring results, no significant changes were observed in either the control or reinforced sections during the first six months following installation. However, the instrumented RPPs started moving after a rainfall incident in September 2011. The instrumented RPP in the control section (IM5) experienced a few increments and smaller reductions in strain compared to the instrumented RPPs in Reinforced Sections 1 and 2 (IM3 and IM8). The higher strain observed in the member IM5 represented greater movement in the slope in the unreinforced control zone. In addition, no significant increment in strain was observed in either of the RPPs installed in the reinforced sections, which signifies that very little movement occurred in these.

The top seven rows of RPPs were installed at 3 ft c/c spacing in Reinforced Section 1. The spacing of the RPPs in the middle portion of Reinforced Section 1 was 6 ft c/c. RPP IM3 was installed in the middle portion of the slope. Figure 7.15 indicates that there was no change in strain at IM3, compared to IM5 and IM8. There is a possibility that the 3 ft c/c spacing at the crest of the slope provided significant resistance and prevented initiation of slope movement. RPP IM8 was installed in the middle section of Reinforced Section 2, which had a larger spacing at the crest of the slope of 4 ft c/c. The strain at IM8 increased slightly during the initial period of rainfall in November 2011, which indicated that there might be slight movement at that location in Reinforced Section 2. Therefore, based on the monitoring results, it could be concluded that 3 ft c/c spacing at the crest of the slope provided better reinforcement than 4 ft c/c spacing.

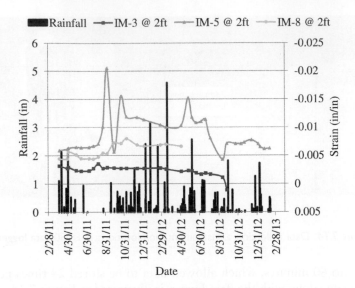

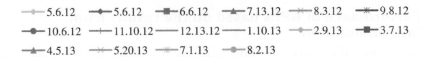

Figure 7.15 Comparison of strain between IM3, IM5 and IM8.

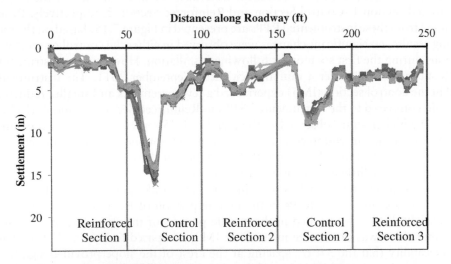

Figure 7.16 Total settlement along the crest of US 287 slope.

The total settlement over the crest of the slope was measured during each survey and is plotted in Figure 7.16. The plot shows that the control sections of the slope had significant settlement at the crest when compared to the reinforced sections. The maximum settlements in Control Sections 1 and 2 were 15 in and 9 in, respectively. It should be noted that Reinforced Section 1 had the lowest spacing of RPPs

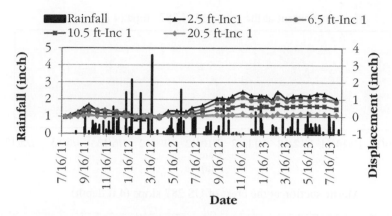

Figure 7.17 Variation in horizontal displacement at Inclinometer 1 in Reinforced Section 1.

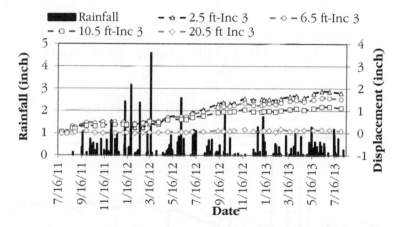

Figure 7.18 Variation in horizontal displacement at Inclinometer 3 in Reinforced Section 2.

(3 ft c/c) at the crest of the slope. Reinforced Sections 2 and 3 had 4 ft c/c spacing at the crest, which was longer than Reinforced Section 1. The instrumented RPP reported almost no movement in Reinforced Section 1, apparently as a result of the closer spacing at the crest of the slope. The total settlement plot shows that the lowest vertical movement (2 in) was in Reinforced Section 1, and this was in good agreement with the results from the instrumented RPPs. On the other hand, the relatively higher vertical movement (3 in) of Reinforced Section 2 could be attributed to the propagation effect of the crack. Since control sections were placed at both sides of Reinforced Section 2, the significant settlement of these sections may have propagated through Reinforced Section 2 and resulted in more settlement there.

The field monitoring results from Inclinometer 1 are presented in Figure 7.17. Inclinometer 1 had increasing horizontal displacement during July to September 2011, the initial period after installation. However, after October 2011 the movement of the slope reduced. The slope showed similar cyclic behaviour between July and September

Moisture content at the crest of US 287 slope (4 ft depth)

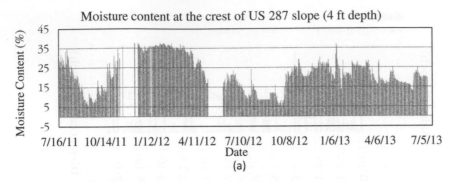

(a)

Martic suction at the crest of US 287 slope (4 ft depth)

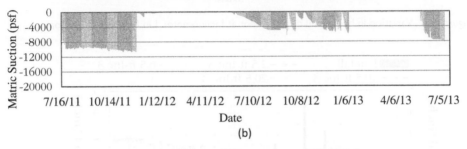

(b)

Comparison of horizontal movement at US 287 slope

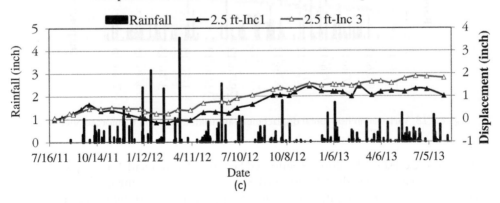

(c)

Figure 7.19 Instrumentation results at the US 287 Slope, a. variation of moisture content at crest, b. variation of matric suction and c. comparison of horizontal displacement at 2.5 feet depth.

2012, and an increase in horizontal displacement again took place. The cyclic displacement behaviour could have occurred due to shrinkage and swelling behaviour in the high-plasticity clayey soil. The maximum movement of the slope, as observed from the inclinometer results, was 1.3 in, which was observed near the surface of the slope. In addition, with the increasing depth of the slope, the horizontal movement of the slope dropped until a depth of 20.5 ft was reached, where almost no movement took place.

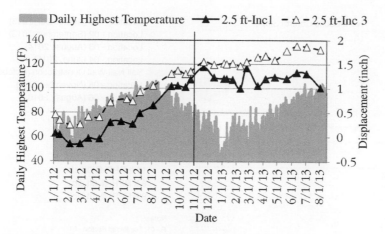

Figure 7.20 Variation in horizontal displacement with daily highest temperature.

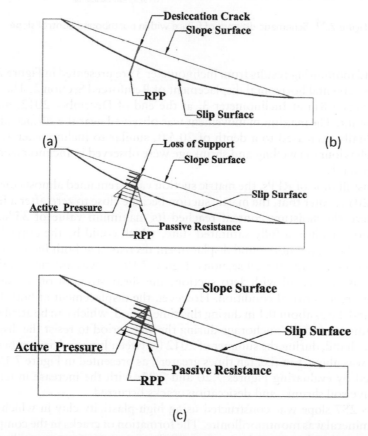

Figure 7.21 Schematic of loss of support for RPP during a dry period: a. formation of desiccation crack during summer; b. passive resistance reduced due to loss of support; c. desiccation crack disappears during wet period and RPP regains passive resistance.

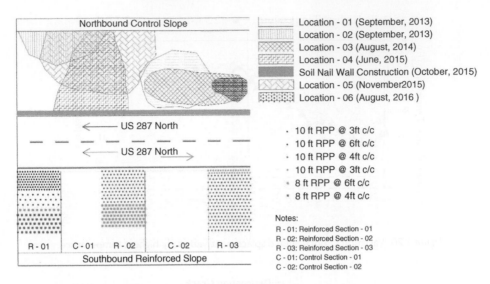

Figure 7.22 Schematic diagram of failure within northbound control slope.

The field monitoring results from Inclinometer 3 are presented in Figure 7.18. They depict an incremental horizontal displacement of Reinforced Section 2. The maximum movement was 1.8 in at Inclinometer 3, at the end of December 2012, and became almost constant. The maximum movement was observed near the surface of the slope, and it gradually decreased to a depth of 20.5 ft, similar to Inclinometer 1. However, no cyclic behaviour of swelling and shrinkage were observed in Inclinometer 3, similar to Inclinometer 1.

After installation of RPPs, the matric suction value remained almost constant up to December 2011. After that, the matric suction became close to zero after a few rainfall events, where the moisture content reached its maximum value of 35%. The low matric suction signifies a fully saturated state, which could be the critical condition for the slope. However, horizontal displacement decreased at both Inclinometer 1 and Inclinometer 3 in the reinforced sections (Figure 7.19). It was anticipated that during the high moisture period with zero suction, the shear strength of the slope would decrease and reach critical condition. However, the displacement at both Reinforced Sections 1 and 2 was about 0.1 in during this time period, which can be attributed to the RPPs providing sufficient anchorage during the wet period to resist the displacement. On the other hand, during the summer of 2012 (between June and September), the RPP movement was about 1 in (i.e. ten times greater), as presented in Figure 7.19. This can be explained by evaluating Figures 7.20 and 7.21: with the increase in temperature, the expansive soil shrunk, and desiccation cracks occurred.

The US 287 slope was constructed using high-plasticity clay in which the major dominant mineral was montmorillonite. The formation of cracks in the compacted clay largely depends on the mineral content, and the presence of montmorillonite makes the soil highly susceptible to the formation of desiccation cracks (Inci, 2008). The desiccation cracks created a loss of support to the RPPs, as illustrated in Figure 7.21,

Location–01 (September 2013)   Location–02 (September 2013)

Location–03 (August 2014)   Location–04 (June 2015)

Location–05 (November 2015)   Location–06 (August 2016)

Figure 7.23 Failure locations on the northbound slope of US 287 (Khan et al., 2016).

reducing the passive resistance of the RPPs and resulting in greater deformation during the summer period. The desiccation cracks disappeared during the wet period, and the RPPs regained their full passive resistance. It can be concluded that the reinforced sections performed better than the unreinforced control sections. In addition, the spacing of the RPPs played an important role in resisting crest settlement and displacement.

7.1.7 Performance of the unreinforced northbound slope

The northbound slope of highway US 287 has been inspected visually on a monthly basis for the past five years, and a number of shallow slope failures were recorded on this control slope. The locations of the failures are presented in Figure 7.22, and photographs of the failures at Locations 1 to 6 are shown in Figure 7.23.

After the installation of RPPs on the southbound slope, the first-time failures (Location 1 and Location 2) on the northbound control slope were observed during September 2013, following a rainfall event. As these were shallow failures with no significant damage to the landscape, no maintenance effort was considered. During August 2014, the slope failed again (Location 3) after a period of heavy rainfall. Soil was backfilled at the failure location, which was a temporary solution to prevent movement of the slope. A global failure (Location 4) of the slope was observed in June 2015 after a significant rainfall of 125 mm during that month. A soil nail wall was constructed to control the movement of the pavement, and the slope was reconstructed in October 2015, when the highly plastic clay soil was stabilised with lime. Despite this reconstruction, the slope failed again (Location 5) in November 2015 after a month of heavy rainfall. The most recent failure was recorded in August 2016 and labelled as Location 6.

During the failures of the northbound slope, a minor increment in horizontal displacement was observed at Inclinometer 1 on the southbound slope; however, no sign of failure was observed at the reinforced section. Moreover, no failure was reported at the control section of the southbound slope. It should be noted that the control sections are located between the reinforced sections. During the failure of the northbound slope, the control section should tend to slip. However, it might get resistance from the adjacent reinforced section and restricted any failure at the control section of the control section of southbound slope.

7.2 HIGHWAY SLOPE NEAR MOCKINGBIRD LANE, DALLAS, TEXAS

This slope was located along I35E (North), near the Mockingbird Lane overpass in Dallas, TX. A crack was observed over the shoulder due to surficial movement at the slope, as illustrated in Figure 7.24. The crack propagated up to 42 ft over the shoulder. The schematic of the failure condition is shown in Figure 7.25. Investigation of the Mockingbird site was performed in February 2014.

7.2.1 Site investigation

Two 2D RI tests were conducted at the site, and the imaging sections identified as RI-1 and RI-2. RI-1 was located over the top of the slope, and RI-2 was located near the toe of the slope. The tests were performed using 56 electrodes that were spaced 5 ft apart. The 2D RI line setups and field investigation results are presented in Figures 7.26 and 7.27.

The resistivity profile for RI-1 indicated that there was a high-resistivity zone at the surface of the slope in the failure area. The depth of the high-resistivity zone was 7 ft. It should be noted that during the surficial soil movement, the failure zone might have become loose and/or disturbed, causing the high resistivity. Therefore, the depth of failure due to the surficial soil movement could be as much as 7 ft. A low-resistivity zone

Figure 7.24 Shoulder crack and surficial movement of the Mockingbird slope.

was observed immediately below the failure zone, which might signify the presence of a high-moisture zone beneath the failure area.

Three soil test borings were conducted in February 2014 on the Mockingbird slope. One soil boring was located over the crest of the slope, and the other two were located near the toe of the slope. The depth of each test boring was 30 ft. Both disturbed and undisturbed samples were collected from the borings. Results from the soil test borings and laboratory tests are shown in Figure 7.28. The presence of a high-moisture zone was observed between a depth of 10 to 30 ft, based on the laboratory investigation of samples taken from the boring at the crest of the slope. The high-moisture zone from this soil boring was in good agreement with the resistivity profile below the failure zone. The soil boring results indicated that the soil type was high-plasticity clayey

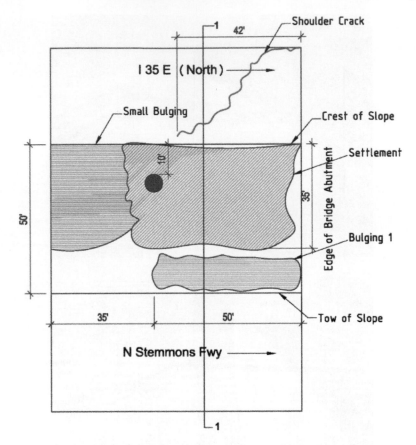

Figure 7.25 Schematic of failure of the Mockingbird slope.

soil, which usually experiences softening of shear strength within the first few years following construction due to its shrinkage and swelling behaviour. Moreover, the shrinkage cracks act as a conduit for rainwater intrusion and possible saturation of the top of the slope, which might cause the failure of the slope.

7.2.2 Slope stability analysis and design of slope stabilisation

Slope stability analyses were performed using the FEM program, PLAXIS (Plaxis, 2011). Two different soil layers were identified during field investigation, and modelling was performed using the fully softened shear strength of the soil for the top soil. The soil strength beyond the failure zone was not reduced for the analysis. The soil parameters used for the analyses are shown in Table 7.4. The slope stability analyses indicated that the FS of the slope was 1.03, which is very close to failure. The failure plane is illustrated in Figure 7.29.

RPPs were proposed for reinforcing the slope, and a further slope stability analysis was conducted, considering 4 in × 4 in rectangular fibre-reinforced RPPs, similar to

Figure 7.26 Resistivity imaging field setup at Mockingbird slope: a. line RI-1; b. line RI-2.

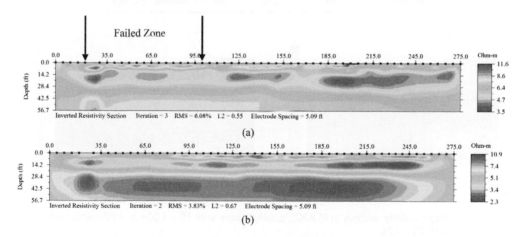

Figure 7.27 Resistivity imaging results at Mockingbird slope: a. RI-1; b. RI-2.

those used at US 287. The length of the proposed RPPs was 10 ft. In addition, six rows of RPPs near the crest of the slope were proposed to have a c/c spacing of 3 ft; for the rest of the slope, near the toe, a 5 ft c/c spacing was proposed. The slope stability analysis was continued to evaluate the FS of the reinforced slope, which was found to be 1.74. The failure plane of the reinforced slope is shown in Figure 7.29b. The proposed layout of the RPPs for the Mockingbird slope is presented in Figure 7.30.

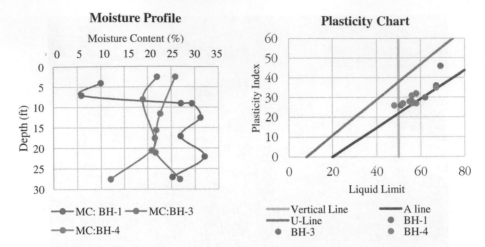

Figure 7.28 Moisture content and plasticity of soil samples from Mockingbird slope.

Table 7.4 Soil parameters used for the slope at Mockingbird slope.

Material	γ (lb/ft³)	c (psf)	φ	E (lb/ft²)	ν
Top soil	125	40	17	15500	0.35
Bottom soil	130	250	20	200000	0.3

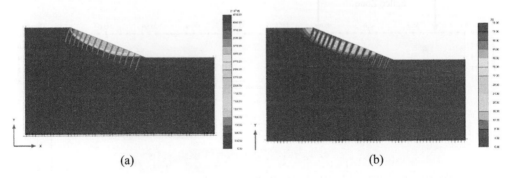

(a)   (b)

Figure 7.29 Slope stability analysis in PLAXIS: a. failed slope with FS = 1.034; b. RPP-reinforced slope with FS = 1.74.

7.2.3 Field installation

Field installation began in May 2014. A crawler-mounted rig with pseudo-vibratory hammer (model: Casagrande M9-1) was used to install the RPPs. However, the steepness of the slope at the crest meant that the rig was not suitable, and there was a high possibility that it might tilt during the installation operation; the work was stopped because of this safety concern. Instead, the installation work resumed with an excavator equipped with a hydraulic breaker (model: Deere 200D with FRD F22

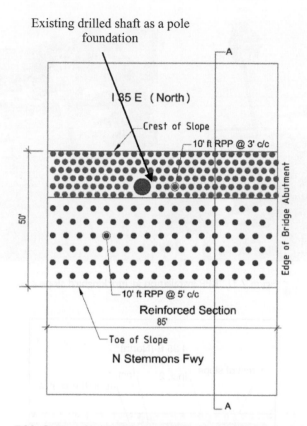

Existing drilled shaft as a pole foundation

I 35 E (North)

Crest of Slope

10' ft RPP @ 3' c/c

50'

Edge of Bridge Abutment

10' ft RPP @ 5' c/c

Reinforced Section
85'

Toe of Slope

N Stemmons Fwy

Figure 7.30 Section details of slope stabilisation on Mockingbird slope.

hydraulic hammer). The excavator performed well in terms of safety and pin installation, which continued for two days. The Mockingbird slope had had repetitive failures and, previously, the slope had been stabilised with lime during maintenance operations. This meant that some locations of the slope had very stiff soil, making it difficult to install or drive the RPPs into it. As a result, some RPPs buckled or broke during the installation. A total of 130 RPPs were installed at the Mockingbird slope during this first phase.

The second phase of RPP installation was carried out in October 2014. A Caterpillar CAT 320D LRR rig with a CAT H130S hydraulic hammer was used to install the RPPs at I-35. The pin installation was carried out in the midst of a serious sustained drought. Hence, at the time of the pin installation, soil conditions were extremely dry and stiff-to-hard. Two masking moulds were prepared for mounting on the hydraulic hammer, with iron chains or straps to connect either the iron pins or the RPPs. One mould was welded with an iron nail having the same cross-sectional area of the RPPs. This was hammered into the ground so that the RPPs could be installed easily, as shown in Figure 7.31. During this second phase of installation, 121 RPPs were installed, making a total of 251 RPPs installed to stabilise the Mockingbird site during Phases 1 and 2.

Figure 7.31 RPP installation at Mockingbird slope.

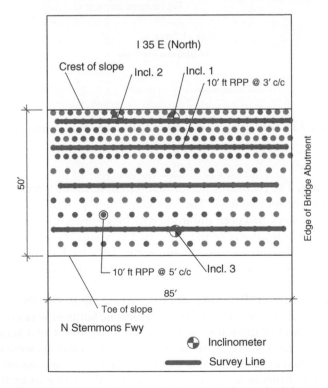

Figure 7.32 Field instrumentation layout of Mockingbird slope.

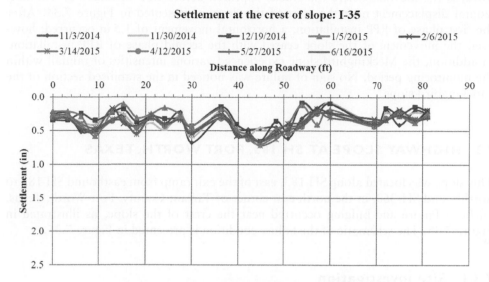

Figure 7.33 Settlement at the crest of Mockingbird slope.

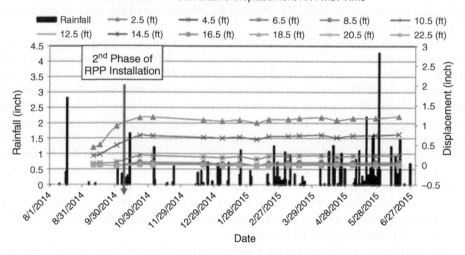

Figure 7.34 Cumulative horizontal displacement plot at Mockingbird slope.

7.2.4 Field instrumentation and performance monitoring

To evaluate the performance of the Mockingbird slope, inclinometers were installed at the crest and toe of the slope and were monitored on a bi-weekly basis. A topographic survey was conducted on a monthly basis. The layout of the inclinometer and the topographic survey is shown in Figure 7.32.

The total settlement plot, as represented in Figure 7.33, showed that there was no significant settlement at the crest. The maximum settlement was about 0.7 inches towards the location where the cracks appeared before RPP installation. The horizontal displacement near the crest of the slope is presented in Figure 7.34. After the first phase of RPP installation, a horizontal movement of 1.5 in occurred; however, the movement of the slope ceased with the second phase of RPP installation. In addition, the Mockingbird slope experienced various intensities of rainfall within the monitoring period. No sign of failure was noticed in the stabilised section of the Mockingbird slope.

7.3 HIGHWAY SLOPE AT SH 183, FORT WORTH, TEXAS

This slope was located along SH 183, east of the exit ramp from eastbound SH 183 to northbound SH 360, in the north-east corner of Tarrant County, Fort Worth, Texas. Surficial failure and bulging occurred near the crest of the slope, as illustrated in Figure 7.35. The schematic of the failure condition is presented in Figure 7.36.

7.3.1 Site investigation

An extensive site investigation was undertaken at the SH 183 slope. It included soil sampling from test borings, laboratory testing of the soil samples collected, and a geophysical investigation using 2D electrical RI. The layout of the soil test borings and RI lines are presented in Figure 7.37.

The RI investigations were conducted using an 8-channel AGI SuperSting resistivity meter with 56 electrodes. The electrodes were placed 5 ft apart during each RI test to develop a profile for each line. The 2D profiles of the RI tests are shown in Figure 7.38.

Based on the 2D RI profiles, a 7 ft deep high-resistivity zone was observed near the crest of the slope, which might have occurred due to disturbance during slope failure or the presence of an active zone. It should be noted that the resistivity of the soil depends on the soil type, moisture conditions and void ratio of the soil. Due to slope movement, the soils in the failed zone become loose, which may result in a higher void ratio and high resistivity. In contrast, the presence of moisture results in low resistivity of the soil. As depicted in Figure 7.38, a high-resistivity zone existed along the tension crack, which may indicate the extent of the tension crack, above a low-resistivity zone that might be a moist zone due to rainwater intrusion through the tension crack. The tension crack extended up to 10 ft in depth.

The liquid limit and the plasticity index of the soil samples from each borehole were determined and are presented in Figure 7.39. Based on the soil test results, the soil samples were classified as low- to high-plasticity clay, according to the USCS. Soil in the top layer, from boreholes BH-1 and BH-2, was classified as low-plasticity clay (CL).

The moisture content of the soil samples collected from borehole BH-1 ranged from 20% to 22%. During the field visit, shoulder cracks were observed near the crest of the slope, which could serve as a potential pathway for the intrusion of rainwater.

The soil samples from different depths within each borehole were further investigated to evaluate the peak, fully softened and residual shear strengths of the soil. The peak shear strength of the undisturbed samples is summarised in Table 7.5. The peak fully softened shear strength and residual shear strength of the remoulded samples are

Figure 7.35 Failure of the SH 183 slope.

presented in Table 7.6. The results of the shear strength tests were used to conduct slope stability analysis and the design of the slope stabilisation scheme.

7.3.2 Slope stability analysis and design of slope stabilisation scheme

Slope stability analyses were performed using the FEM program PLAXIS and the Mohr–Coulomb soil model. Modelling was performed using the fully softened shear strength of the soil for the top 7 ft of soil, because a possible softened zone (disturbance)

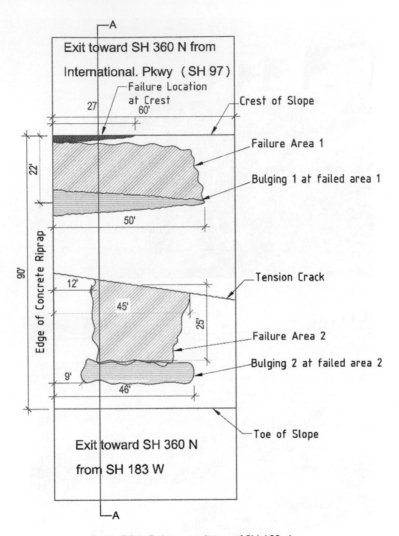

Figure 7.36 Failure condition of SH 183 slope.

was observed at a depth of 7 ft during RI. To simulate the fully softened layer, a new layer was introduced during the analysis. The fully softened shear strength from the test results was utilised, considering the wet-dry cycles.

For the analysis, the soil strength beyond the failure zone was considered to be peak shear strength. The soil slope model and soil properties are presented in Figure 7.40 and Table 7.7, respectively. Without the consideration of any water table, the FS obtained from the analyses was found to be 1.46. The analyses were continued, incorporating a perched water zone (a fully moistened zone) due to rainfall. It was assumed that due to the seasonal wet-dry cycles, the high-plasticity CH clay contracted and created a path for the intrusion of rainwater. This rainwater may have saturated the top few

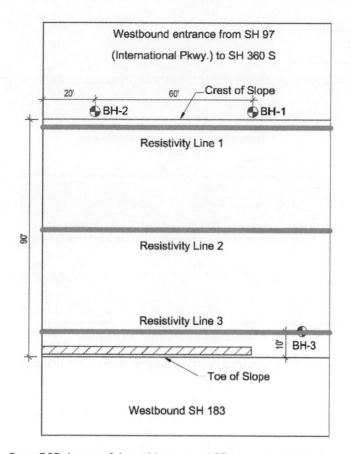

Figure 7.37 Layout of the soil borings and 2D resistivity imaging lines.

feet of the soil and remained there for several days due to the low permeability of the high-plasticity clay soil, forming a temporary perched water zone. During this analysis, the temporary perched water zone (fully moistened zone) was considered to be at a depth of 5 ft. The slope stability analysis indicated that the FS reduced to 1.09 in the presence of such a zone at the top of the slope.

A slope stabilisation plan was designed, using 4 in × 4 in × 10 ft fibre-reinforced RPPs. The section and type of RPP were selected according to the field performance of the reinforced US 287 slope (Section 7.1). Various spacings of the RPPs were used in trial runs, and the FS of the reinforced slope was determined. Based on several iterations in PLAXIS, a 3 ft c/c spacing of the RPPs, in both horizontal directions, was selected near the crest of the slope, followed by a 4 ft c/c spacing throughout the rest of the slope. The outputs of the slope stability analyses for the reinforced areas of the slope are presented in Figure 7.41. The FS of the slope increased to 1.667 with 25 layers of RPPs. Based on the safety analysis, the slope stabilisation plan for the SH 183 slope, using RPPs, is presented in Figure 7.42.

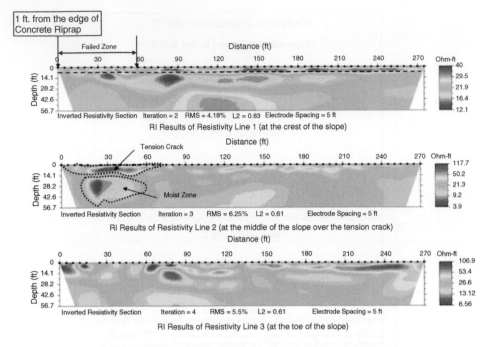

Figure 7.38 2D electrical resistivity imaging results at SH 183 slope.

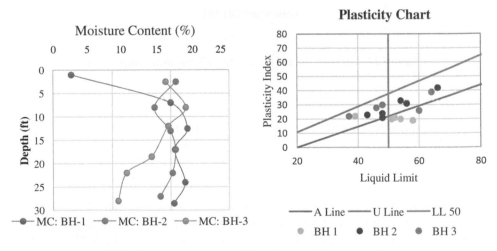

Figure 7.39 Moisture content profiles and plasticities for each borehole.

7.3.3 Field installation

A 60 ft × 90 ft section of the slope was stabilised using 10 ft long RPPs placed 3 ft apart c/c in the upper 30 ft of the slope. The remainder of the slope had an RPP spacing of

Table 7.5 Summary of shear strength tests on undisturbed samples.

Borehole no.	Sample depth (ft)	Specimen	Test type	Cohesion (psf)	Friction angle
BH-1	3	Undisturbed	Direct shear	550	21
BH-2	19	Undisturbed	Direct shear	500	13
BH-3	3	Undisturbed	Direct shear	403	20
BH-3	15	Undisturbed	Consolidated-undrained triaxial	610	12
BH-3	19	Undisturbed	Direct shear	950	16

Table 7.6 Summary of shear strength tests on remoulded samples.

Borehole no.	Sample depth (ft)	Specimen	Test type	Cohesion (psf)	Friction angle
BH-3	3	Remoulded with wet-dry cycle	Direct shear	120	21
BH-3	3	Remoulded	Ring shear	142	26

Table 7.7 Parameters for FEM analysis.

Soil type —	Friction angle φ (°)	Cohesion c (psf)	Unit weight Γ (pcf)	Elastic modulus E (ksf)	Poisson ratio ν —	RPP properties Parameter	Unit	Value
1	20	120	125	100	0.35	E	ksi	308
2	21	400	19.6	100	0.3	A	in²	12.25
3	20	950	16	200	0.25	I	in⁴	12.5

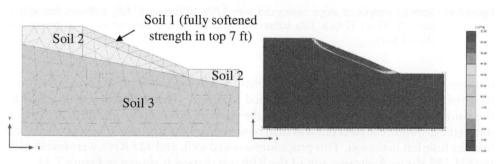

Figure 7.40 Slope stability analysis of the SH 183 slope: a. soil model; b. FS = 1.098, considering perched water at 5 ft.

4 ft c/c. All of the RPPs installed at the SH 183 site were 10 ft long. The RPPs were installed during September 2014, after a very hot, dry summer in Texas. Hence, at the time of the RPP installation, soil conditions were extremely dry and very stiff-to-hard in consistency. The regular method of driving RPPs into the ground resulted

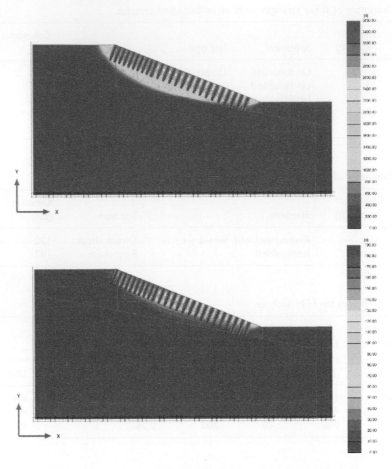

Figure 7.41 Stability analysis of slope reinforced with RPPs 10 ft long: a. fully softened strength in top 7 ft (FS = 1.754); b. fully softened soil up to 7 ft and perched water up to top 5 ft (FS = 1.667).

in them breaking. To overcome this problem, a full-sized steel pin was manufactured to drive holes into the required/predefined locations. The steel pin was attached to the hammer and used to make a hole in the stiff soil layer, up to 7 ft deep, by driving it into the ground. The steel pin was then removed and the RPP was immediately driven into the hole left in the soil. This procedure worked well, and 425 RPPs were installed in the SH 183 slope. A photograph of the RPP installation is shown in Figure 7.43.

There was a section of the slope (see Figure 7.44) where it proved very difficult to install RPPs, which caused delays during the installation. It was discovered that there was a retaining wall at that location, which was part of a previous slope stabilisation exercise. However, even though the slope had been stabilised with a retaining wall, part of the slope above and part of the slope below the retaining wall had failed. It was apparent, therefore, that some slopes in North Texas have failed even after being repaired with a range of expensive methods.

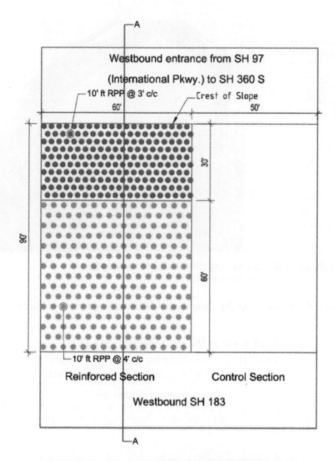

Figure 7.42   Proposed RPP layout for SH 183 slope.

Figure 7.43   Installation of RPP at SH 183 slope.

Installation of RPPs was very difficult due to existing buried
retaining structure

Figure 7.44 Location of the zone where RPP installation was unsuccessful.

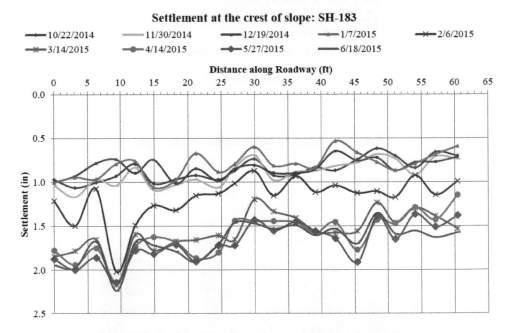

Figure 7.45 Settlement profile at the crest of the reinforced slope.

7.3.4 Field instrumentation and performance monitoring

A topographic survey was conducted on a monthly basis to observe the movement
of the RPPs at the SH 183 slope. The total settlement over the crest of the slope was

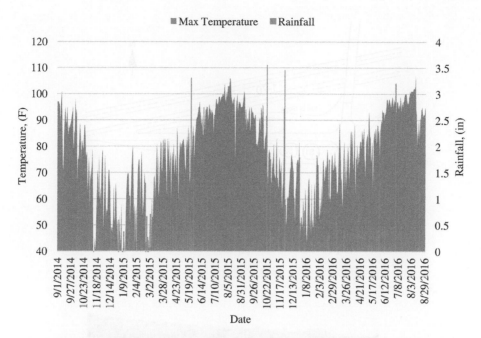

Figure 7.46 Maximum temperature and rainfall distributions within five miles of SH 183 slope (at Dallas/Fort Worth Airport, Dallas, Texas) following RPP installation.

measured during each survey, and the plotted result is shown in Figure 7.45. The plot showed no significant incremental settlement at the crest. The maximum settlement was about 2.2 in, towards the abutment of the bridge.

The SH 183 slope was constructed over highly plastic clay soil that is subjected to wetting and drying activity due to seasonal moisture and temperature variations. Khan et al. (2016) conducted a detailed failure investigation of the highway slope, which indicated that, during the summer, the soil dried out and had desiccation cracks. The desiccation cracks significantly increased the vertical permeability of the soil and acted as a preferential flow path for rainwater to enter the slope and saturate the soil. As a result, the matric suction of the soil disappeared and reduced the FS of the slope. After installation of the RPPs, there was heavy rainfall during late spring (May 2015) and early fall (October 2015), when the maximum temperature was higher than 80°F, as presented in Figure 7.46. Due to the heavy rainfall in the summer, when highly plastic soils experienced shrinkage/desiccation cracks, several slope failures occurred in other slopes near the RPP-reinforced SH 183 slope. However, no sign of failure or movement of soil was noticed in the RPP-stabilised slope at SH 183.

7.4 INTERSTATE 70 (I-70) EMMA FIELD TEST SITE IN COLUMBIA, MISSOURI (LOEHR AND BOWDERS, 2007)

The I-70 Emma field test site is located on I-70, approximately 65 miles west of Columbia, Missouri. The height of the slope is 22 ft, with 2.5:1 (horizontal: vertical)

(a)

(b)

Figure 7.47 a. Location of the slide areas in I-70 site; b. south side of embankment at I-70 Emma site with slide area S1 at left, S2 at centre and S3 at right (Loehr and Bowders, 2007).

side slopes that form the eastbound entrance ramp to I-70 in Saline County. The slope is composed of mixed lean clays, with scattered cobbles and construction rubble (concrete and asphalt). The slope has experienced recurring slides in four areas of the embankment over the past decade or more. The plan view of the slide areas of the slope, denoted as S1, S2, S3 and S4, is shown in Figure 7.47.

7.4.1 Site investigation

A total of 11 borings from slides S1 and S2 were made across the site to depths ranging from 10 to 33 ft (3 to 10 m). Continuous 3 in (7.6 cm) diameter Shelby tube samples were collected from each boring and were labelled for further testing. The collected soil samples indicated that the slope is composed of a mixture of lean (CL) and fat (CH) clays, with scattered fine gravel and cobbles. Groundwater was present near the slide areas of S1 and S2, below the elevation of the toe of the slope. Standard penetration tests (SPTs) performed near the base of the embankment produced N60 values between 3 and 8. Two other SPTs performed below the base of the embankment produced N60s in excess of 40. Moisture content was observed to be essentially constant with depth throughout the embankment and ranged from 14% to 34%, with the vast majority of values lying between 20% and 25%. Atterberg limits determined for samples from the site indicated that the soils have liquid limits (LLs) between 39 and 60, plastic limits (PLs) between 19 and 27, and plasticity indices (PIs) between 10 and 41.

Loehr and Bowder (2007) determined the Mohr–Coulomb effective stress shear strength parameters from both triaxial compression and direct shear tests. The stress path determined from consolidated-undrained (CU, or R-test) and consolidated-drained (CD, or S-test) triaxial compression tests, along with upper bound and lower bound failure envelopes established from the test results for the surficial soils and soils at greater depths, are presented in Figure 7.48 and Table 7.8. Based on the test results, the effective stress cohesion intercept, c, for the surficial soil is equal to approximately 100 psf (4.8 kPa), and the effective stress friction angle, φ, is equal to 23°. For the deeper soils, c ranges from 170 to 365 psf (8.1 to 17.5 kPa), and φ is approximately 25°.

7.4.2 Slope stabilisation scheme

The failed slope at the Emma site was stabilised in two separate phases. During Phase I, the stabilisation of slide areas S1 and S2 was addressed. Initially, slide area S3 was utilised as a control slope to monitor the performance of this Phase I stabilisation. Later on, the failure area S3 was stabilised in Phase II. The reinforcement configurations selected for stabilisation of slide areas S1, S2 and S3 are presented in Figure 7.49. RPPs were placed 3 ft apart c/c in a staggered grid in both S1 and S2 areas. In addition, while the RPPs in slide area S1 were placed perpendicular to the face of the slope, those in slide area S2 were installed vertically. The FS for both of these slope stabilisation schemes was estimated to be approximately 1.2, based on calculations performed using the back-calculated soil conditions.

Slide area S3 was divided into four sections, denoted as Sections A through D, with a different reinforcement scheme utilised in each section. Thus, in Section A, members were placed on a 4.5 ft (longitudinal) ×3.0 ft (transverse) staggered grid. A 4.5 ft × 6.0 ft grid was used in Section B, a 6.0 ft × 6.0 ft grid in Section C, and a 6.0 ft × 4.5 ft grid in Section D. The FS of the different sections of slide area S3 was calculated using two different conditions, which are summarised in Table 7.9. In condition A, the back-calculated soil parameters for an FS equal to unity were used. A two-layer profile, with a perched water condition within the upper layer, was assumed, as presented in Figure 7.50. For condition B, the upper layer was assumed to have $c = 95$ psf (4.5 kPa) and $\varphi = 15°$, while the lower layer had $c = 310$ psf (14.8 kPa) and $\varphi = 22°$. In addition, the piezometric line for the upper layer was assumed to be at the ground surface. The FS of the unreinforced slope, in relation to condition B, was also 1.0. With the slope stabilisation scheme, the FS of slide S3 was marginally increased, ranging between 1.02 and 1.16.

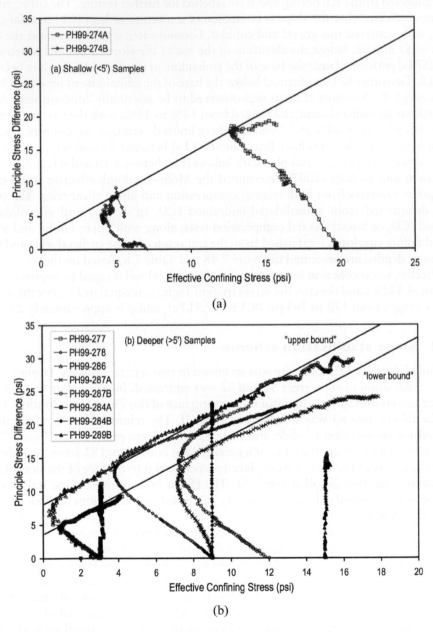

Figure 7.48 Summary of triaxial test results for specimens from the I-70 Emma site: a. shallow samples; b. deeper samples (Loehr and Bowders, 2007).

7.4.3   Field installation

The pin installations at slides S1 and S2 were conducted during November and December 1999. In slide area S1, reinforcing members were installed approximately

Table 7.8 Mohr–Coulomb effective stress parameters from I-70 Emma Site (Loehr and Bowders, 2007).

Soil layer	Depth	Upper bound		Lower bound		Direct shear	
		c (psf)	φ (°)	c (psf)	φ (°)	c (psf)	φ (°)
Surficial clay	<4 ft	96	23	–	–	202	14
Deeper clay	>4 ft	364	25	170	25	101	14

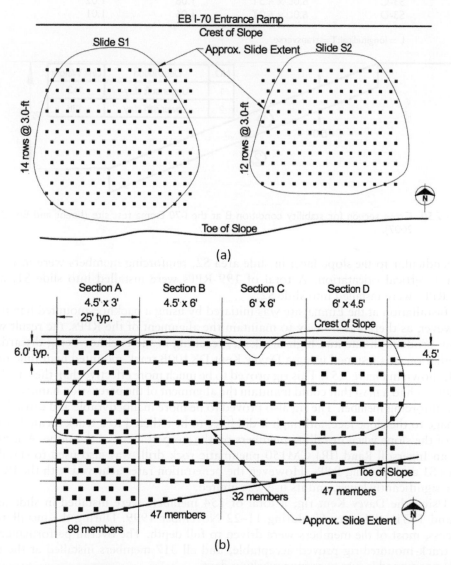

Figure 7.49 Layout of RPPs in the slide area of I-70 Emma site: a. slide areas S1 and S2; b. slide area S3 (Parra et al., 2003).

Table 7.9 Calculated factors of safety for slide areas S1, S2 and S3 (Loehr and Bowders, 2007).

Slope section	RPP spacing (ft)	Factor of safety	
		Condition A	Condition B
S1 and S2	3.0L × 3.0T	1.20	1.21
S3-A	4.5L × 3.0T	1.16	1.10
S3-B	4.5L × 6.0T	1.10	1.03
S3-C	6.0L × 4.5T	1.08	1.02
S3-D	6.0L × 6.0T	1.06	1.01

L = longitudinal; T = transverse.

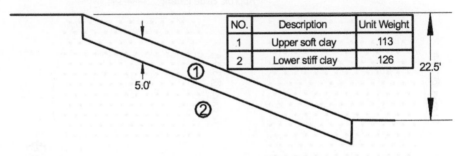

NO.	Description	Unit Weight
1	Upper soft clay	113
2	Lower stiff clay	126

Figure 7.50 Cross section for stability condition B at the I-70 Emma test site (Loehr and Bowders, 2007).

perpendicular to the slope face; in slide area S2, reinforcing members were installed with a vertical orientation. A total of 199 RPPs were installed into slide S1, and 163 RPPs were installed into slide S2.

Installation at the Emma site was initiated by using a backhoe-mounted hammer. However, as there was no way to maintain the alignment of the RPPs, the result was a low production rate and high breakage, resulting in the backhoe being discarded. Installation was resumed using a Davey Kent DK100B track-mounted hydraulic rock drill, shown in Figure 7.51. This rig proved to be much more effective than the backhoe because it had a mast that could maintain the alignment of the percussion hammer with the reinforcing member. The rig also proved to be more manoeuvrable and caused less damage to the slope face, although it did have to be tethered to a truck located at the top of the slope when installing members on the steepest areas of the slope. A second rig, an Ingersoll Rand (IR) CM150 pneumatic rock drill, was also used to stabilise slides S1 and S2 using RPPs. However, the penetration rates achieved with the IR rig were significantly lower, and its use was limited.

Using the Davey Kent rig, a total of 154 members were installed in slide area S1 and 163 in slide area S2 during 11–22 November 1999. During the installation process, most of the members were driven to full depth. The overall performance of the track-mounted rig proved acceptable, and all 317 members installed at the site were positioned in just over four working days.

Field installation in slide area S3 at the I-70 Emma site took place on 6–7 January 2003. Before the slope stabilisation, the sliding area was regraded to the original slope

Figure 7.51   Davey Kent DK100B track-mounted hydraulic rig at I-70 Emma site (Loehr and Bowders, 2007).

configuration. RPPs were installed using two different pieces of equipment: an Ingersoll Rand ECM350 track-mounted drill rig, with an extended boom and a simple drop-weight device, and the Farm King post driver, commonly used for driving fence or guard-rail posts, mounted on a skid-steer loader. Both types of equipment performed exceptionally well, with a rapid production rate. A total of 199 reinforcing members were installed in slide area S3. RPPs were generally driven without any significant problems, and the overall installation was completed in less than two working days.

The penetration rates during these installations are summarised in Table 7.10. It should be noted that the penetration rate considers only the time required to drive an RPP into the ground. Average penetration rates varied from 0.4 ft /min to 10.2 ft/min. In addition, at peak installation rate, around 80–100 RPPs could be installed in a single working day.

7.4.4 Instrumentation and performance monitoring

After installation, inclinometers and instrumented RPPs were installed to monitor the performance of the stabilised slides. In slide areas S1 and S2, the installed instrumentation included instrumented members to monitor loads on the reinforcing members, slope inclinometers to monitor lateral deformations in the slope, continuously screened wells to monitor possible positive pore water pressures, and "jet-filled" tensiometers to monitor possible soil suction. A total of ten members were instrumented, using 120-ohm electrical resistance strain gauges, and installed in slide areas S1, S2 and S3 during Phase I (slide area S3 was utilised as a control slope during this phase). During Phase I, five slope inclinometers were installed to monitor deformation of the slope. The inclinometer casings were extended approximately 5 ft below the toe of the slope.

Table 7.10 Penetration rate for RPPs at the I-70 Emma site (Loehr and Bowders, 2007).

Stabilised slope	RPP length	Penetration rate (ft/min)			
		Min.	*Max.*	*Avg.*	*Std. dev.*
S1	8 ft	0.7	10.2	5.0	2.2
	<8 ft	0.7	2.7	1.6	0.7
	All	0.7	10.2	4.6	2.4
S2	8 ft	1.5	8.7	4.5	1.6
	<8 ft	0.4	7.0	2.4	1.4
	All	0.4	8.7	3.9	1.8
S3	8 ft	2.0	18.5	10.1	4.4
	<8 ft	0.1	17.0	4.1	2.8
	All	0.1	18.5	6.5	4.6
	All*	1.2	15.0	4.2	2.9

*using drop-weight hammer driving machine.

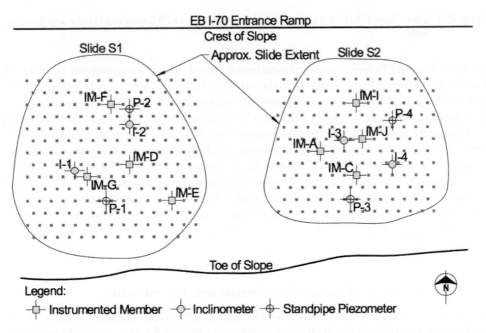

Figure 7.52 Instrumentation layout in slide areas S1 and S2 during Phase I stabilisation at I-70 Emma site (Loehr and Bowders, 2007).

In addition, five continuously screened wells and jet-filled tensiometers were installed at the site. The instrumentation layout for Phase I is shown in Figure 7.52.

During Phase II, additional instrumentation was installed in slide area S3, to monitor the performance of the newly stabilised slope section. The instrumentation scheme included six instrumented RPPs, four inclinometer casings, two clusters of standpipe

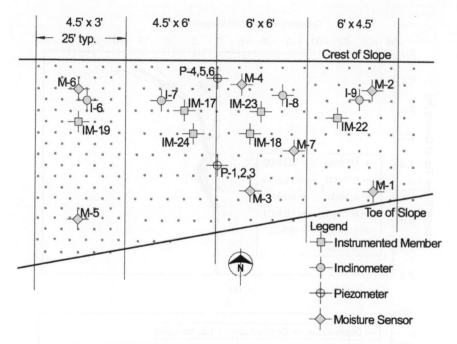

Figure 7.53 Instrumentation layout at the slide area S3 slope stabilised using RPPs in Phase II at I-70 Emma site (Loehr and Bowders, 2007).

piezometers, an array of moisture sensors, and seven profile probe access tubes. The layout of the instrumentation in slide area S3 is shown in Figure 7.53. The inclinometer casings were installed in each of the four reinforced sections and were extended 19 ft below ground level to provide adequate length for anchorage.

Based on the field monitoring results from inclinometer I-2 (in slide area S1), the cumulative displacement vs. depth, and cumulative displacement vs. time are presented in Figure 7.54. It was observed that movements were generally minimal for the first year following installation; then, the movements increased to a maximum of approximately 0.8 in over the next 6 months. Movements were minimal thereafter. Parra et al. (2003) observed that the movements corresponded closely with the rainfall data from the site. Both control slides (S3 and S4) failed in late spring of 2001, when small movements were observed in the stabilised areas. In addition, the continuously screened piezometers and tensiometers installed at the site indicated that increased pore water pressures were present during spring of 2001.

The maximum bending moments determined from the strain gauges on three instrumented RPPs are presented in Figure 7.55. The field monitoring results indicated that the bending moment of member IM-C, which was installed in slide S2, gradually increased between installation and April 2001, followed by a small but rapid increase in measured bending moment in May 2001, around the time of control section failure. Member IM-G in slide S1 showed a relatively large initial increase in bending moment over the first six months after installation, followed by an essentially constant

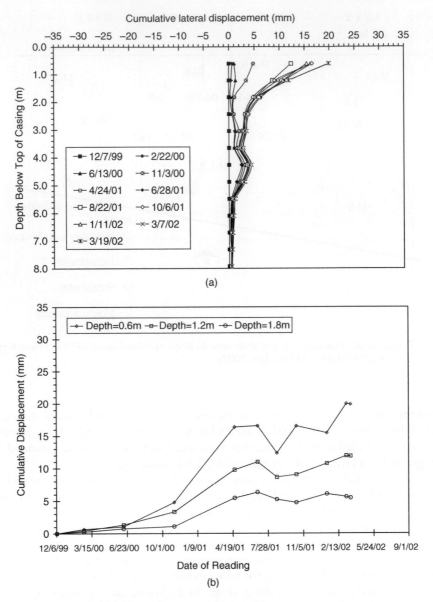

Figure 7.54 Inclinometer I-2 data from slide area S1 at I-70 Emma site: a. displacement vs. depth; b. displacement vs. time (Parra et al., 2003).

maximum bending moment over time before being damaged by mowing operations in May 2001. In the control slide, instrumented member IM-H showed behaviour similar to that observed for IM-C, except that the bending moments increased more dramatically in the months leading up to the control section failure. At the time of the failure, IM-H indicated a bending moment of approximately 850 lbs/ft, a value that is very

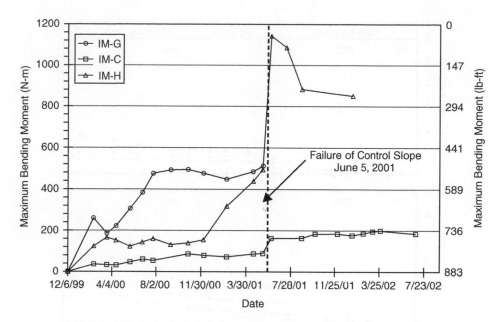

Figure 7.55 Bending moment diagram from instrumented RPPs at I-70 Emma site (Parra et al., 2003).

near the average moment capacity, 900 lbs/ft, of the reinforcing members. Member IM-H failed in bending terms when the control slide failed.

The horizontal displacement over time for each of the four sections of slide area S3 is presented in Figure 7.56. Corresponding periods of low, medium and high precipitation are also shown at the top of the figure. Figure 7.56 indicates that all inclinometers exhibited a similar pattern of displacement over time, which is also the case for other monitored field demonstration sites. A little displacement was observed during the first six months after installation, followed by an increase in displacement that was presumed to be the result of decreased stability and the mobilisation of resistance in the reinforcing members. There was a 4.5-inch rainfall event between 31 August and 1 September 2003, which probably reduced the FS of the slope. During this period, mobilisation of the load occurred for a period of time, with an increase in displacement at a small but relatively constant rate.

From September 2004 onwards, an additional increase in displacements was observed in Sections B and C of slide area S3. This second period of movement is attributed to additional mobilisation of resistance from the reinforcing members. Moreover, high precipitation was observed during late summer and fall 2004. The piezometer installed in the field indicated an increase in the perched water table due to the high precipitation. Subsequently, a failure occurred in Sections B and C between November 2004 and January 2005. A plan and photograph of the failure are shown in Figure 7.57. The failure was approximately 25–30 feet wide, and was confined to the upper portion of the slope, stopping approximately 10 feet above the toe of the slope in Sections B and C. The failure of Sections B and C was attributable to the wider spacing, 6 ft, of their RPPs, which failed to provide resistance to the sliding.

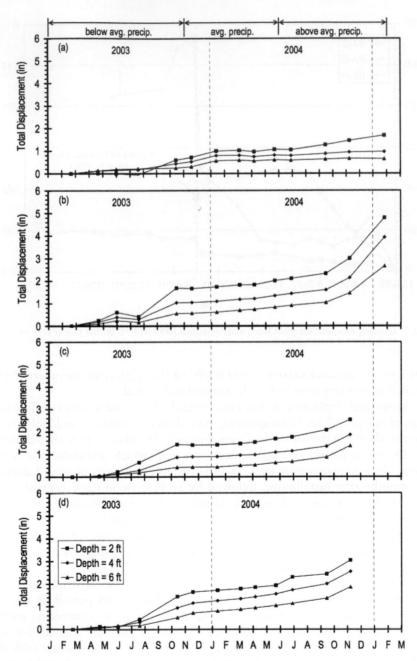

Figure 7.56 Displacement profile of slide area S3: a. Section A; b. Section B; c. Section C; d. Section D (Loehr and Bowders, 2007).

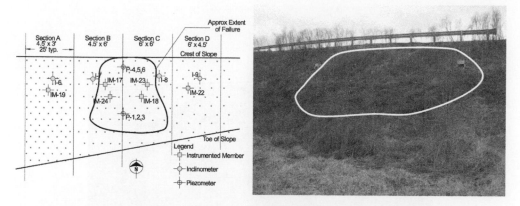

Figure 7.57 Approximate extent of failure at slide area S3 (Loehr and Bowders, 2007).

It is notable that the deformations measured in Section A were significantly smaller than those observed in the other three sections (see Figure 7.56). A slight increase in deformation was observed between August and October 2003, in response to increased precipitation and load mobilisation in the RPPs. When the adjoining sections failed in late 2004, no such increase in deformation was observed in Section A, due to its more closely spaced RPPs (3–4.5 ft).

7.5 INTERSTATE 435 (I-435)–WORNALL ROAD FIELD TEST SITE, MISSOURI

The I-435 Wornall Road test site is located at the intersection of I-435 and Wornall Road in southern Kansas City, Missouri. It is a zoned-fill embankment consisting of a 3–5 ft surficial layer of mixed lean-to-fat clay, of soft-to-medium consistency, overlying stiffer compacted clay shale. The slope was approximately 31.5 ft high, with side slopes of 2.2:1 (horizontal:vertical) and had experienced at least two surficial slides along the interface between the upper clays. It was chosen for stabilisation using RPPs. A photograph of the slope failure appears in Figure 7.58.

7.5.1 Site investigation

Loehr and Bowders conducted a site investigation of the I-435-Wornall Road site, using soil boring and laboratory testing. A total of seven hollow-stem auger borings were performed, up to 30 ft deep, and continuous Shelby tube samples were collected. An SPT was performed at 1.5 ft intervals, until refusal. Loehr and Bowders (2007) reported that the SPT N_{60} values at the surficial layer ranged from 0 to 2 at a depth of 4 ft (1.2 m), and values at greater depths ranged from 6 to 16. A limestone layer was encountered at a depth of 12 ft from the toe of the slope, and a groundwater condition was evident in the bottom third of the slope.

Based on the laboratory test results, Loehr and Bowders (2007) reported that the soils from the surficial layer had LLs ranging from 38 to 51, and PIs from 16 to 34.

Figure 7.58 Failed slope over I-435 near Wornall Road (Loehr and Bowders, 2007).

The compacted shale present below the surficial layer had LLs ranging from 29 to 76, and PIs from 12 to 51. The authors also conducted CU triaxial compression tests with pore water pressure measurements. Stress paths determined from these tests are plotted in Figure 7.59, along with failure envelopes established for surficial (<4 ft) and deeper (>4 ft) soils. Based on the CU tests, the Mohr–Coulomb strength parameters are summarised in Table 7.11.

7.5.2 Slope stabilisation scheme

Loehr and Bowders (2007) designed the slope stabilisation using RPPs, considering that a relatively thin surficial layer overlaid compacted clay shale, as determined by the borings and samplings at the site. The thickness of the surficial layer varied between 3 ft (0.9 m) and 5 ft (1.5 m) for the different analyses. During the site investigation and laboratory testing, Loehr and Bowders noticed that landscaping mulch had been intermixed with the soil during the slide, which made it difficult to obtain high-quality soil samples for testing. As a result, the laboratory test results for the surficial clay layer were questionable, and the authors conducted a series of back analyses. For the back analyses, the strength parameters of the compacted clay materials were considered as $c = 30$ psf and $\varphi = 27°$. Several combinations of pore water pressure and cohesion were back calculated for the surficial clay layer, to obtain an FS equal to 1.0. The soil profile for the stability analysis is represented in Figure 7.60.

The slope stabilisation scheme of the I-435 Wornall Road site involved a total of 643 RPPs, at 3 ft c/c spacing, where the previous slide had occurred. Additional RPPs were placed on a coarser 3 ft × 6 ft (0.9 m × 1.8 m) grid above the slide area to reduce the potential for future sliding in the upper portion of the slope. The FS for the chosen reinforcement configuration was estimated to be between 1.15 and 1.50. The layout of RPPs is shown in Figure 7.61.

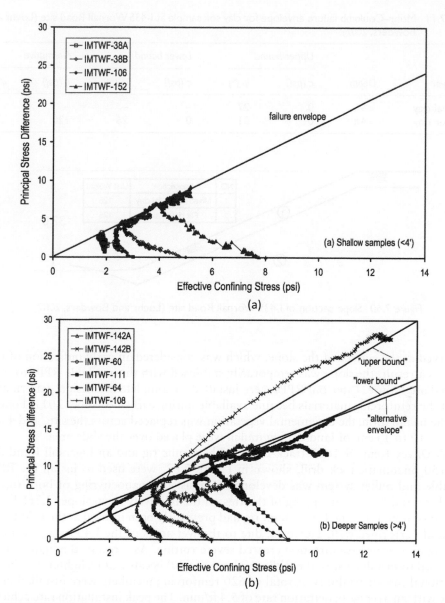

Figure 7.59 Triaxial test results of soil samples from I-435 Wornall Road site at: a. shallow depth; b. deeper depths (Loehr and Bowders, 2007).

7.5.3 Field installation

The field installation of RPPs at the I-435 Wornall Road site began in mid-October 2001. Before the slope stabilisation was started, the slope was regraded to its original slope configuration. Several moderate-to-heavy rainfalls occurred before the installation, which caused seepage in several locations of the slope. In addition, cracks were

Table 7.11 Mohr–Coulomb failure envelope for clay soil sample at I-435 Wornall Road site (Loehr and Bowders, 2007).

Soil layer	Depth	Upper bound		Lower bound		Alternative	
		c (psf)	φ (°)	c (psf)	φ (°)	c (psf)	φ (°)
Surficial clay	<4 ft	0	27	–	–	–	–
Deeper clay	>4 ft	0	31	0	26	120	23

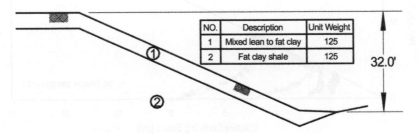

NO.	Description	Unit Weight
1	Mixed lean to fat clay	125
2	Fat clay shale	125

32.0'

Figure 7.60 Slope section of I-435 Wornall Road site (Loehr and Bowders, 2007).

observed at the surface of the slope, which was considered an early indication of failure. As a result, the slope was temporarily stabilised with the available RPPs to resist immediate failure. Later on, RPPs were installed to complete the design pattern after all of the reinforcing materials became available during early December 2011. Following the installation, the ornamental vegetation was replaced across the site, and 4 to 6 inches (10 to 15 cm) of landscaping mulch were placed over the slide area.

A Davey Kent DK100B track-mounted hydraulic rig and an Ingersoll Rand (IR) CM150 pneumatic rock drill, shown in Figure 7.62, were used to install the RPPs. A cable and pulley system was developed to assist the manoeuvring of both rigs on the slope and to prevent tipping of the rigs on the relatively steep slope (2.2H:1V). It should be noted that the Davey Kent rig had previously been utilised at the I-70 Emma site and performed well. However, due to the wet areas of the I-435 site, operating this rig in a transverse direction created severe rutting. As a result, the lighter IR rig was used to install most of the RPPs (590 out of 620) because of its lighter weight and additional manoeuvrability. A total of 620 reinforcing members were installed in the slope with an average penetration rate of 5.4 ft/min. The peak installation rate achieved at the I-435 Wornall Road site was 114 pins per day, with an average installation rate of approximately 80 pins per day. The frequency distribution of average penetration rates for the RPPs is presented in Figure 7.63.

7.5.4 Instrumentation and performance monitoring

Loehr and Bowders (2007) installed instrumentation at the I-435 Wornall Road site in the form of four inclinometers, four instrumented reinforcing members, two clusters of standpipe piezometers, and an array of moisture instrumentation. The instrumented

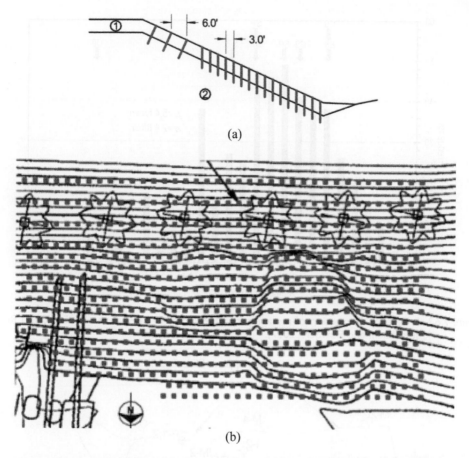

Figure 7.61   Layout of RPPs at the I-435 Wornall Road site: a. cross-sectional view; b. plan view (Loehr and Bowders, 2007).

Figure 7.62   RPP installation at the I-435 Wornall Road site: a. Davey Kent rig; b. Ingersoll Rand rig (Loehr and Bowders, 2007).

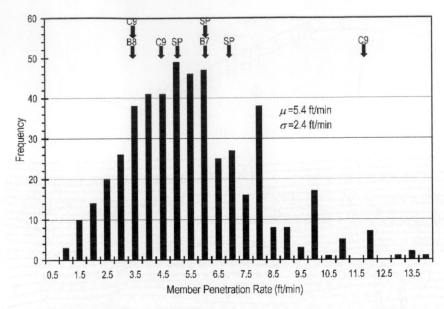

Figure 7.63 Frequency distribution of average penetration rates of RPPs at the I-435 Wornall Road site (Loehr and Bowders, 2007).

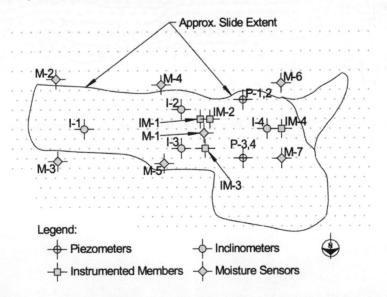

Figure 7.64 Location of instrumentation in I-435 Wornall Road site (Loehr and Bowders, 2007).

reinforcing members were installed as part of the main RPP installation. The first two instrumented members, IM-1 and IM-2, were installed near the centre of the slide; one, IM-3, was placed at the down slope of these previous two; and the fourth instrumented member, IM-4, was installed near the centre where the previous sliding initiated. One inclinometer, I-1, was installed near the eastern side of the stabilised area,

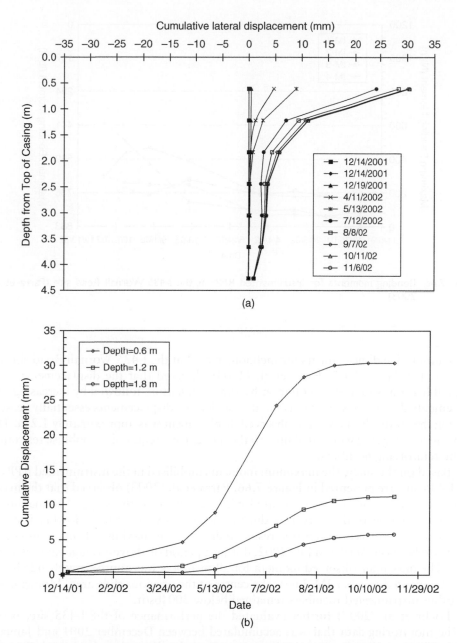

Figure 7.65 Cumulative displacement plots from inclinometer I-2 at the I-435 Wornall Road site: a. displacement vs. depth; b. displacement vs. time (Parra et al., 2003).

and the remaining three inclinometers were installed in proximity to the instrumented members (IM-1 to IM-4). The layout of the instrumentation is shown in Figure 7.64. Approximate depths for the inclinometers were: I-1 – 19.0 ft (5.8 m); I-2 – 26 ft (7.9 m); I-3 – 14.5 ft (4.4 m); I-4 – 19.5 ft (5.9 m).

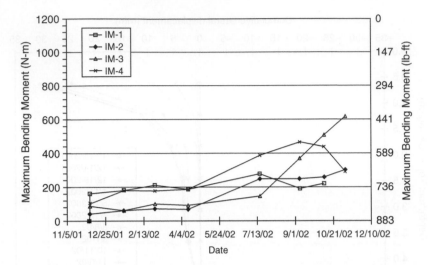

Figure 7.66 Bending moments for instrumented RPPs at the I-435 Wornall Road site (Parra et al., 2003).

Cumulative displacements for inclinometer I-2 at the I-435 Wornall Road site are presented in Figure 7.65. Parra et al. (2003) observed that the displacements were negligible for a period of several months after installation, followed by increasing displacements during the spring season, after which the displacements essentially ceased. During the study, the maximum observed displacement was approximately 1.2 in. The movements observed were attributed to the movements required to mobilise resistance in the reinforcing members.

Based on the study, the maximum moments mobilised in the instrumented RPPs at the I-435 site are presented in Figure 7.66. Parra et al. (2003) observed that the maximum moments were closely correlated with movement in the slope. The maximum moments were generally very low during the first four months following installation. The performance monitoring results indicated that maximum bending moments increased between April and July 2002, during a period of above-average rainfall in the area. The maximum observed moment for all members was approximately 428 lbs/ft, in instrumented member IM-3 near the centre of the previous slide area. Moments for the other instrumented members remained below 300 lbs/ft.

Loehr et al. (2007) further evaluated the performance of the I-435 site, based on the monitoring data that was accumulated between December 2001 and January 2005. The variations in precipitation, piezometric levels, cumulative lateral displacements and mobilised bending moments are presented in Figure 7.67. Loehr et al. (2007) noticed significant deviations from normal trends, including extremely heavy precipitation in May 2002 and precipitation drastically greater than normal in the spring and summer of 2004. In addition, a perched water level was observed in the top 3 to 5 ft, at the lean clay layer near the surface. The piezometric levels were observed to be at their highest during spring and early summer and their lowest during winter.

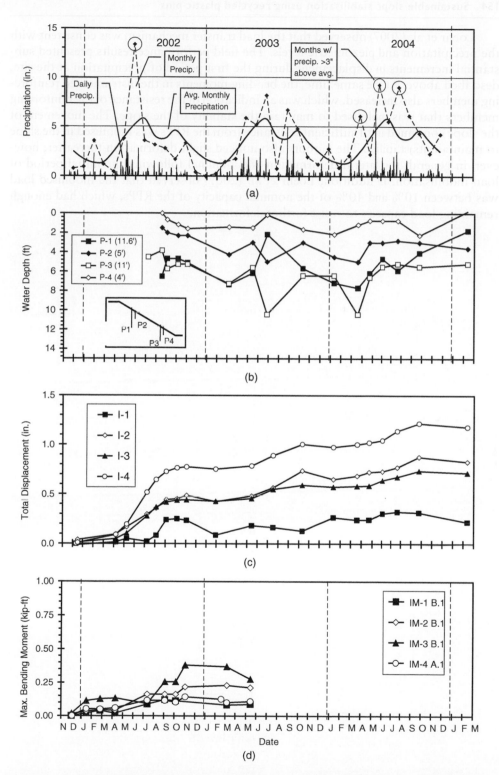

Figure 7.67 Summary of field measurements over time at I-435 Wornall Road site: a. precipitation; b. piezometric levels; c. displacement; d. RPP bending moments (Loehr et al., 2007).

Loehr et al. (2007) observed that the load transfer mechanism was consistent with the precipitation and piezometric levels. The field performance results presented substantial increments in displacement during the first period of precipitation at the site, described above. At the same time, the bending moments in the instrumented reinforcing members also increased, which was an indication of the resistance of the reinforcing members that was mobilised to maintain the stability of the slope. The movement of the slope continued until sufficient resistance from the RPPs was mobilised in the slope to maintain its stability. The authors also noticed slight deformation thereafter; however, in general, the lateral deformation was minimal following the initial period of load mobilisation. In addition, Loehr et al. (2007) observed that the mobilised load was between 10% and 40% of the nominal capacity of the RPPs, which had enough remaining load capacity to resist further deformation.

Appendix A

Design charts

LIST OF FIGURES

(Continued)

Figure no.	c	φ	Slope ratio
A31	300	30	1:3
A32	300	30	1:4
A33	400	0	1:2
A34	400	0	1:3
A35	400	0	1:4
A36	400	10	1:2
A37	400	10	1:3
A38	400	10	1:4
A39	400	20	1:2
A40	400	20	1:3
A41	400	20	1:4
A42	400	30	1:2
A43	400	30	1:3
A44	400	30	1:4
A45	500	0	1:2
A46	500	0	1:3
A47	500	0	1:4
A48	500	10	1:2
A49	500	10	1:3
A50	500	10	1:4
A51	500	20	1:2
A52	500	20	1:3
A53	500	20	1:4
A54	500	30	1:2
A55	500	30	1:3
A56	500	30	1:4

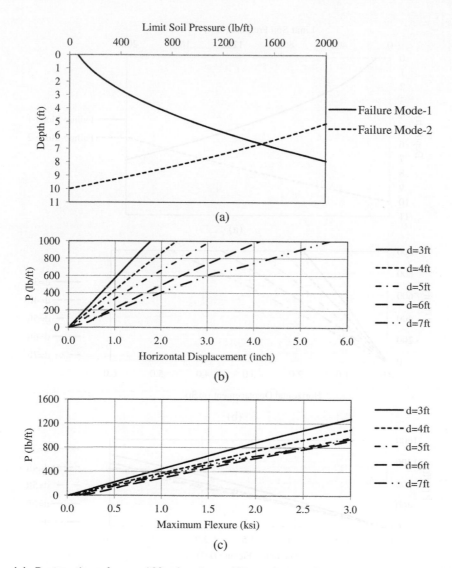

Figure A.1 Design chart for $c = 100$ psf and $\varphi = 10°$: a. limit soil curve; b. load vs horizontal displacement for slope 3H:1V; c. load vs maximum flexure for slope 3H:1V.

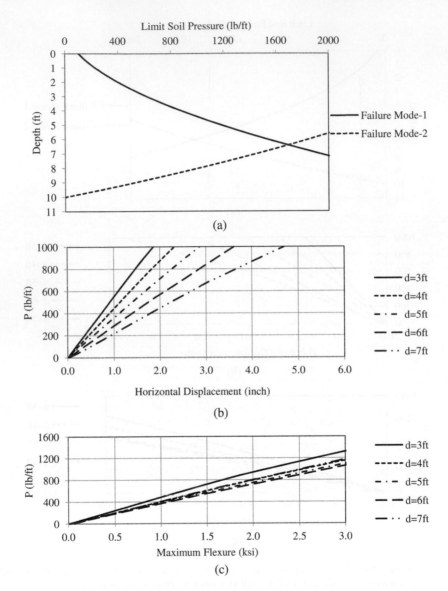

Figure A.2 Design chart for $c = 100$ psf and $\varphi = 10°$: a. limit soil curve; b. load vs horizontal displacement for slope 4H:1V; c. load vs maximum flexure for slope 4H:1V.

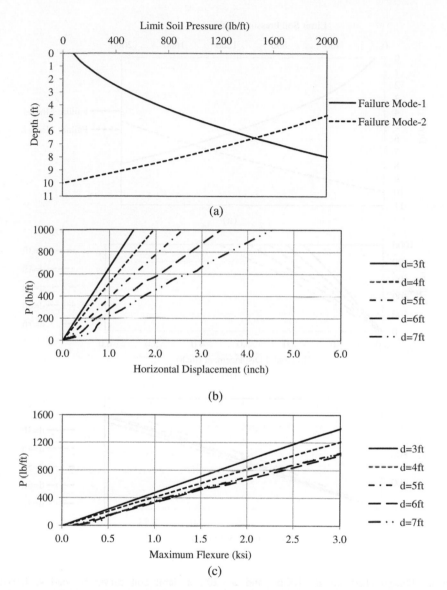

Figure A.3 Design chart for $c = 100\,$psf and $\varphi = 20°$: a. limit soil curve; b. load vs horizontal displacement for slope 2H:1V; c. load vs maximum flexure for slope 2H:1V.

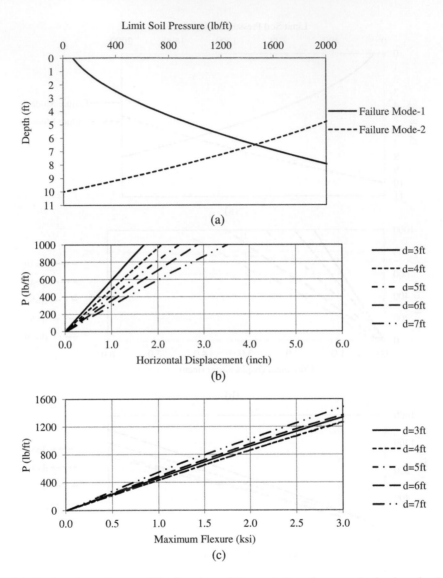

Figure A.4 Design chart for $c = 100$ psf and $\varphi = 20°$: a. limit soil curve; b. load vs horizontal displacement for slope 3H:1V; c. load vs maximum flexure for slope 3H:1V.

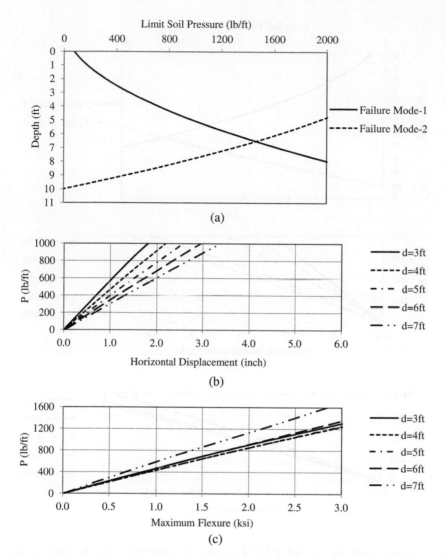

Figure A.5 Design chart for $c = 100$ psf and $\varphi = 20°$: a. limit soil curve; b. load vs horizontal displacement for slope 4H:1V; c. load vs maximum flexure for slope 4H:1V.

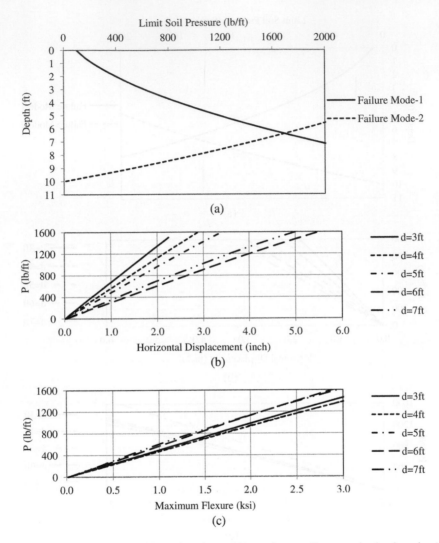

Figure A.6 Design chart for $c = 100$ psf and $\varphi = 30°$: a. limit soil curve; b. load vs horizontal displacement for slope 2H:1V; c. load vs maximum flexure for slope 2H:1V.

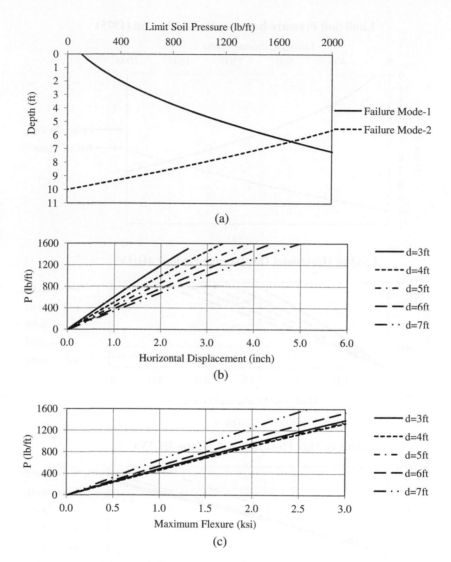

Figure A.7 Design chart for $c = 100$ psf and $\varphi = 30°$: a. limit soil curve; b. load vs horizontal displacement for slope 3H:1V; c. load vs maximum flexure for slope 3H:1V.

Limit Soil Pressure based on Ito and Matsui (1975)

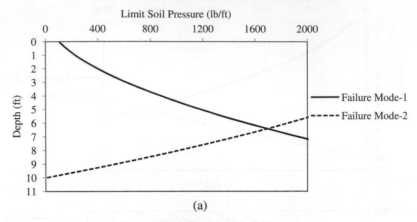

(a)

Load vs Horizontal Displacement (Slope 4H:1V)

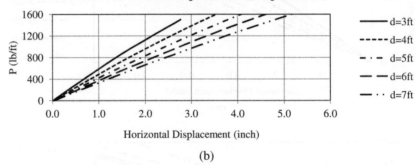

(b)

Load vs Max Flexure Stress (Slope 4H:1V)

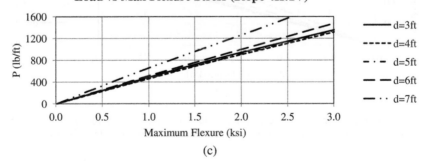

(c)

Figure A.8 Design chart for $c = 100$ psf and $\varphi = 30°$: a. limit soil curve; b. load vs horizontal displacement for slope 4H:1V; c. load vs maximum flexure for slope 4H:1V.

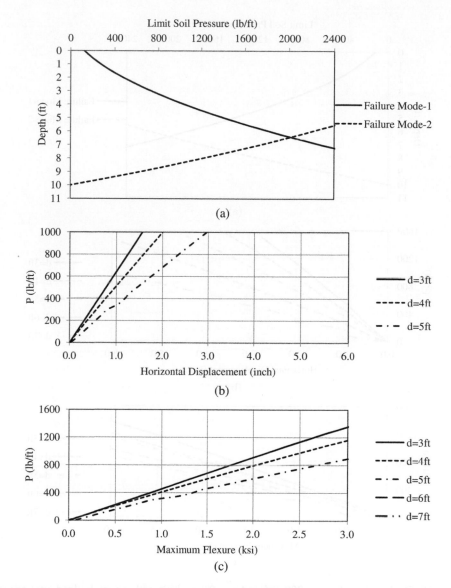

Figure A.9 Design chart for $c = 200$ psf and $\varphi = 0°$: a. limit soil curve; b. load vs horizontal displacement for slope 2H:1V; c. load vs maximum flexure for slope 2H:1V.

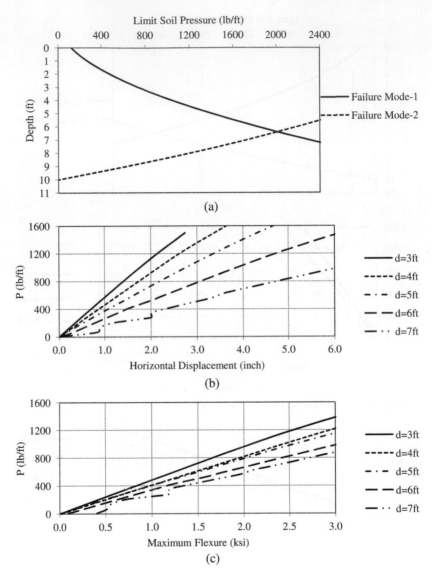

Figure A.10 Design chart for $c = 200$ psf and $\varphi = 0°$: a. limit soil curve; b. load vs horizontal displacement for slope 3H:1V; c. load vs maximum flexure for slope 3H:1V.

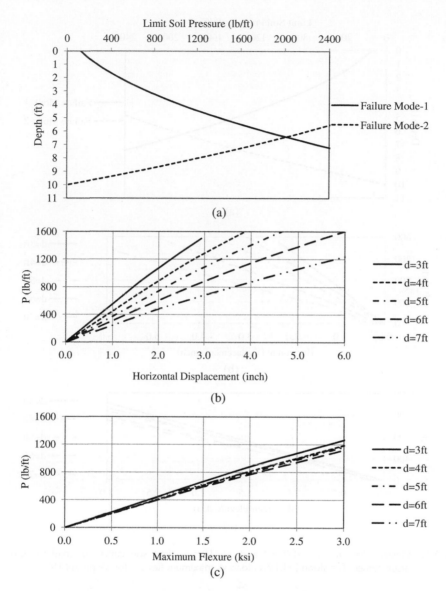

Figure A.11 Design chart for $c = 200$ psf and $\varphi = 0°$: a. limit soil curve; b. load vs horizontal displacement for slope 4H:1V; c. load vs maximum flexure for slope 4H:1V.

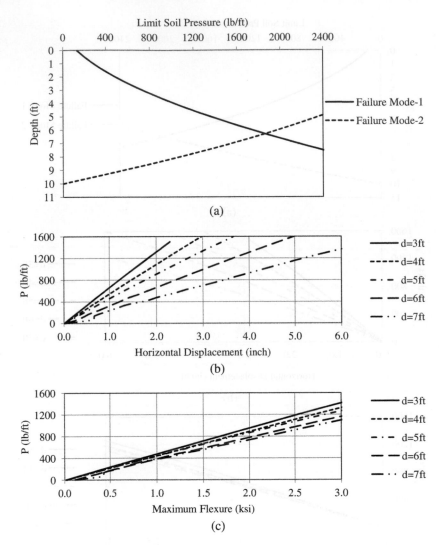

Figure A.12 Design chart for $c = 200$ psf and $\varphi = 10°$: a. limit soil curve; b. load vs horizontal displacement for slope 2H:1V; c. load vs maximum flexure for slope 2H:1V.

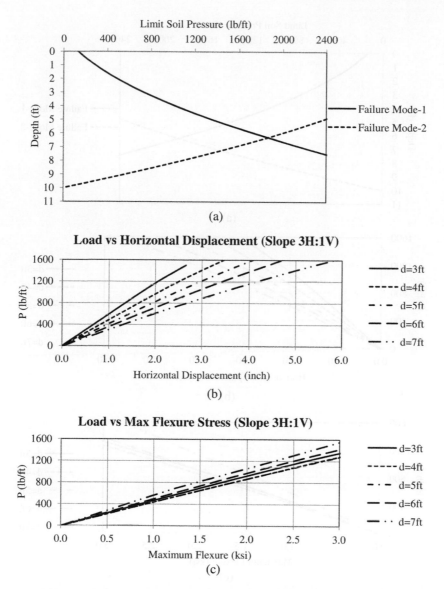

Figure A.13 Design chart for $c = 200$ psf and $\varphi = 10°$: a. limit soil curve; b. load vs horizontal displacement for slope 3H:1V; c. load vs maximum flexure for slope 3H:1V.

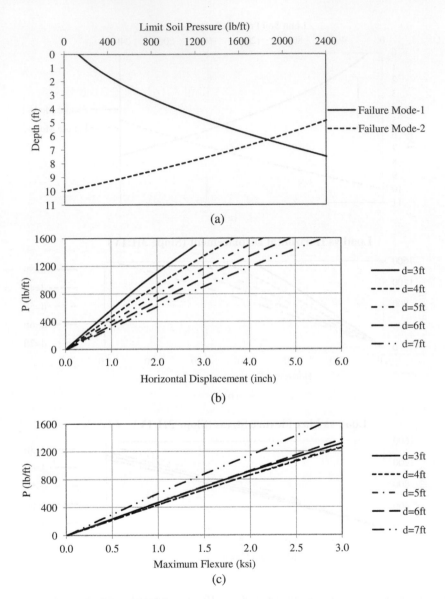

Figure A.14 Design chart for $c = 200$ psf and $\varphi = 10°$: a. limit soil curve; b. load vs horizontal displacement for slope 4H:1V; c. load vs maximum flexure for slope 4H:1V.

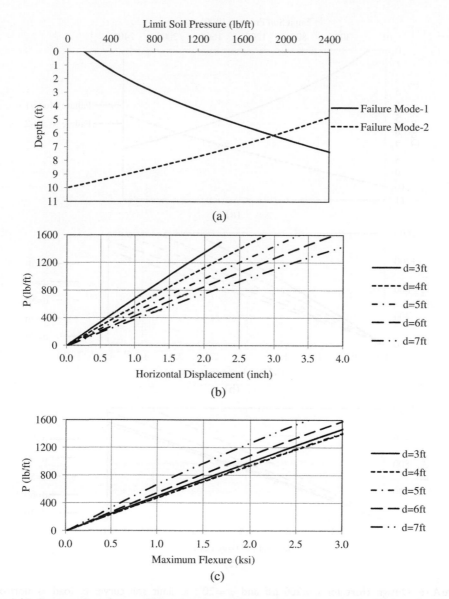

Figure A.15 Design chart for $c = 200$ psf and $\varphi = 20°$: a. limit soil curve; b. load vs horizontal displacement for slope 2H:1V; c. load vs maximum flexure for slope 2H:1V.

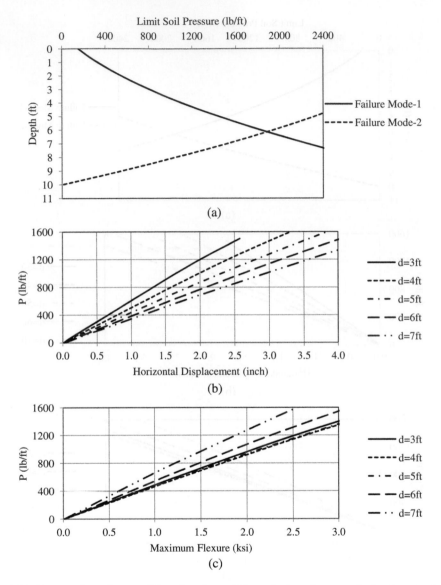

Figure A.16   Design chart for $c = 200$ psf and $\varphi = 20°$: a. limit soil curve; b. load vs horizontal displacement for slope 3H:1V; c. load vs maximum flexure for slope 3H:1V.

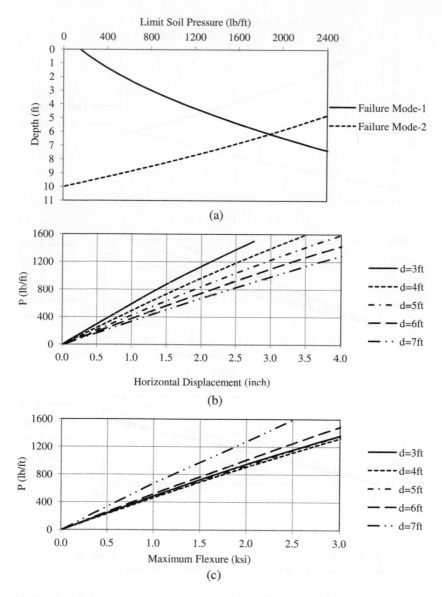

Figure A.17 Design chart for $c = 200$ psf and $\varphi = 20°$: a. limit soil curve; b. load vs horizontal displacement for slope 4H:1V; c. load vs maximum flexure for slope 4H:1V.

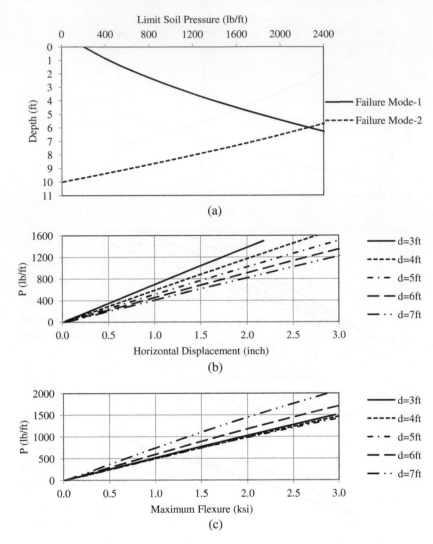

Figure A.18 Design chart for $c = 200$ psf and $\varphi = 30°$: a. limit soil curve; b. load vs horizontal displacement for slope 2H:1V; c. load vs maximum flexure for slope 2H:1V.

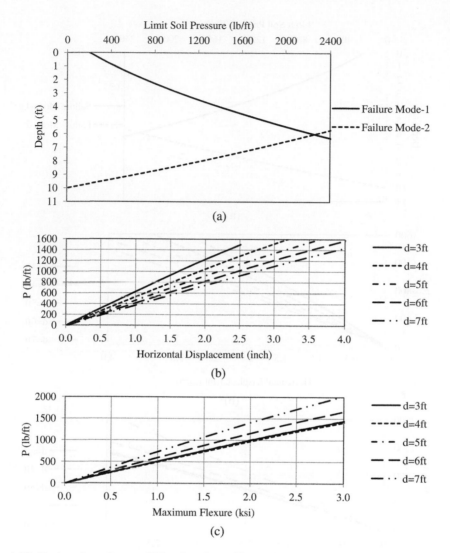

Figure A.19 Design chart for $c = 200$ psf and $\varphi = 30°$: a. limit soil curve; b. load vs horizontal displacement for slope 3H:1V; c. load vs maximum flexure for slope 3H:1V.

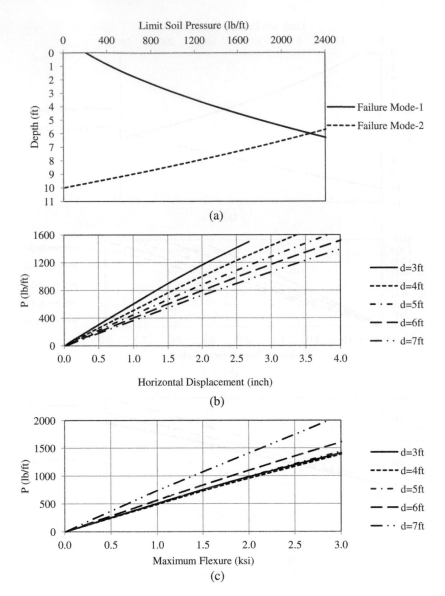

Figure A.20 Design chart for $c = 200$ psf and $\varphi = 30°$: a. limit soil curve; b. load vs horizontal displacement for slope 4H:1V; c. load vs maximum flexure for slope 4H:1V.

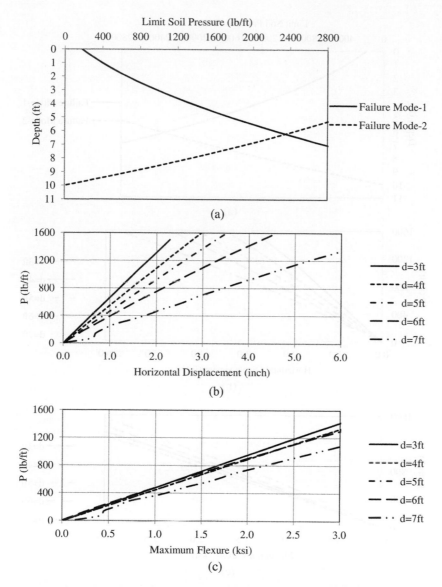

Figure A.21 Design chart for $c = 300$ psf and $\varphi = 0°$: a. limit soil curve; b. load vs horizontal displacement for slope 2H:1V; c. load vs maximum flexure for slope 2H:1V.

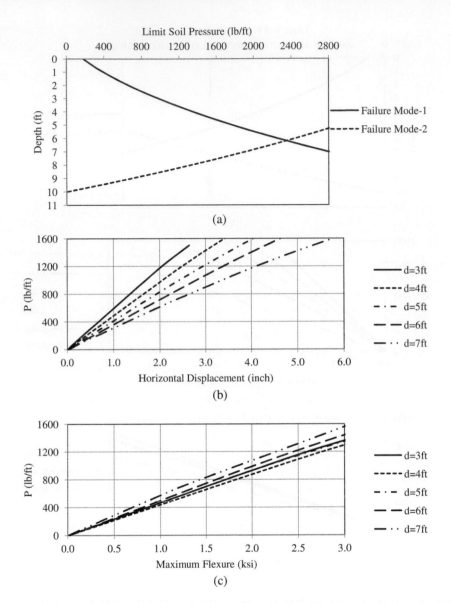

Figure A.22 Design chart for $c = 300$ psf and $\varphi = 0°$: a. limit soil curve; b. load vs horizontal displacement for slope 3H:1V; c. load vs maximum flexure for slope 3H:1V.

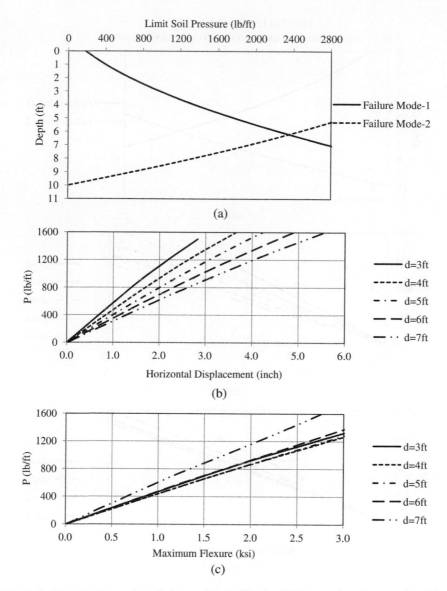

Figure A.23 Design chart for $c = 300$ psf and $\varphi = 0°$: a. limit soil curve; b. load vs horizontal displacement for slope 4H:1V; c. load vs maximum flexure for slope 4H:1V.

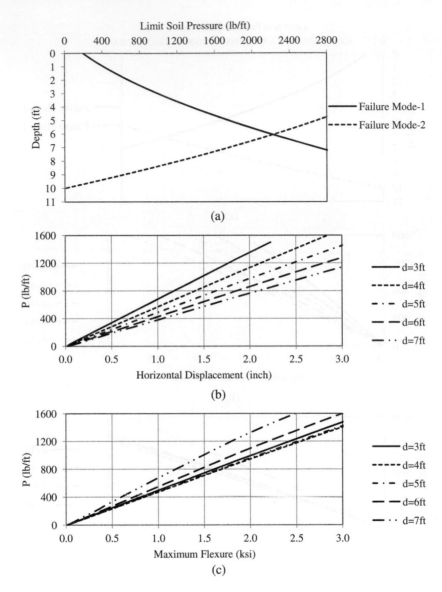

Figure A.24 Design chart for $c = 300$ psf and $\varphi = 10°$: a. limit soil curve; b. load vs horizontal displacement for slope 2H:1V; c. load vs maximum flexure for slope 2H:1V.

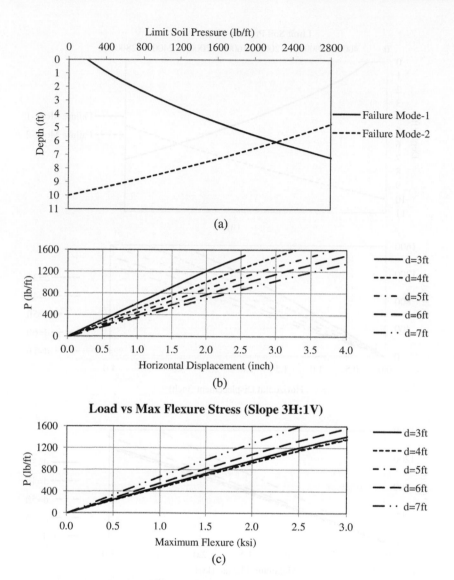

Figure A.25 Design chart for $c = 300$ psf and $\varphi = 10°$: a. limit soil curve; b. load vs horizontal displacement for slope 3H:1V; c. load vs maximum flexure for slope 3H:1V.

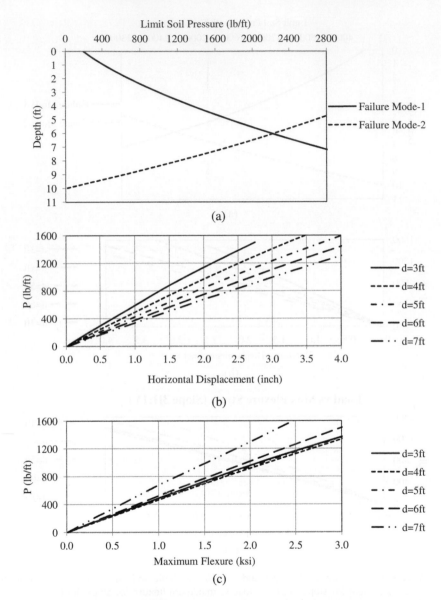

Figure A.26 Design chart for $c = 300$ psf and $\varphi = 10°$: a. limit soil curve; b. load vs horizontal displacement for slope 4H:1V; c. load vs maximum flexure for slope 4H:1V.

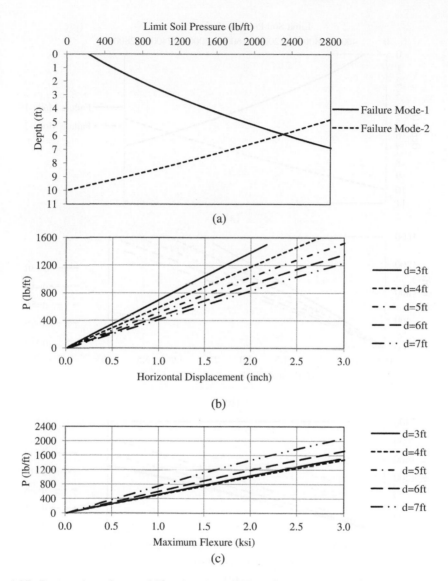

Figure A.27 Design chart for $c = 300$ psf and $\varphi = 20°$: a. limit soil curve; b. load vs horizontal displacement for slope 2H:1V; c. load vs maximum flexure for slope 2H:1V.

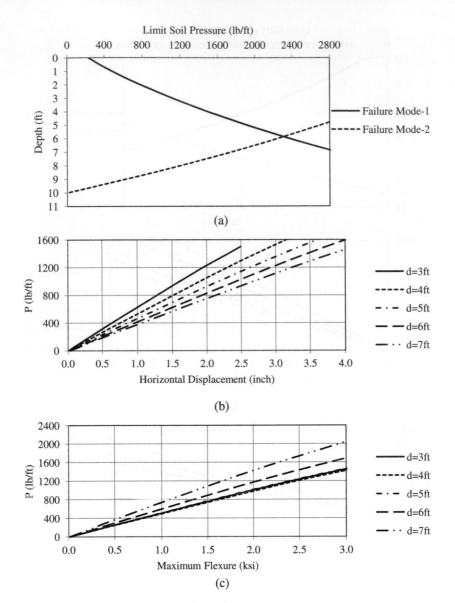

Figure A.28 Design chart for $c = 300$ psf and $\varphi = 20°$: a. limit soil curve; b. load vs horizontal displacement for slope 3H:1V; c. load vs maximum flexure for slope 3H:1V.

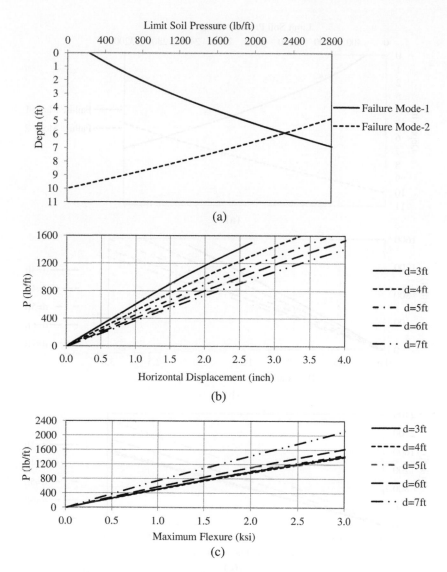

Figure A.29 Design chart for $c = 300$ psf and $\varphi = 20°$: a. limit soil curve; b. load vs horizontal displacement for slope 4H:1V; c. load vs maximum flexure for slope 4H:1V.

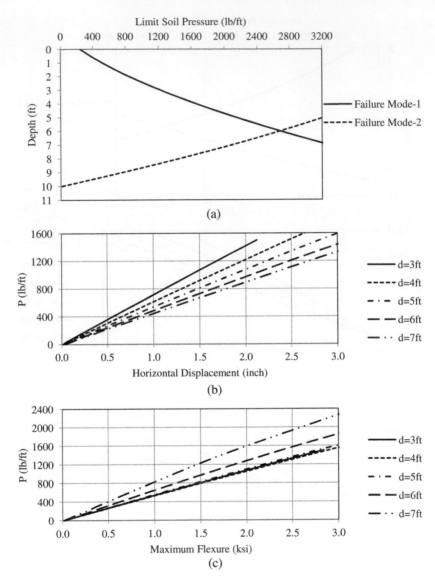

Figure A.30 Design chart for $c = 300$ psf and $\varphi = 30°$: a. limit soil curve; b. load vs horizontal displacement for slope 2H:1V; c. load vs maximum flexure for slope 2H:1V.

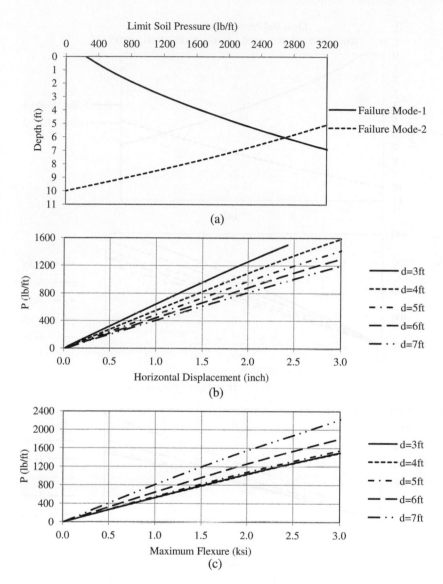

Figure A.31 Design chart for $c = 300$ psf and $\varphi = 30°$: a. limit soil curve; b. load vs horizontal displacement for slope 3H:1V; c. load vs maximum flexure for slope 3H:1V.

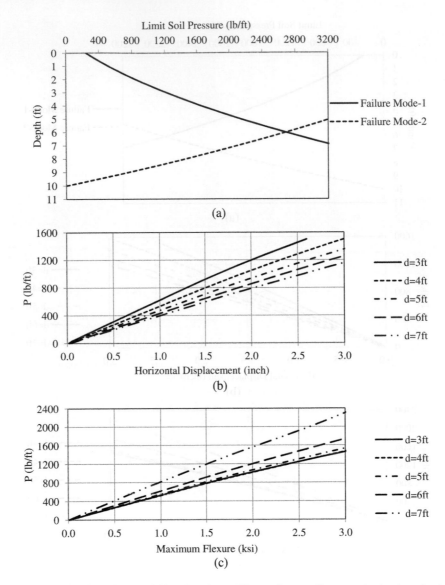

Figure A.32 Design chart for $c = 300$ psf and $\varphi = 30°$: a. limit soil curve; b. load vs horizontal displacement for slope 4H:1V; c. load vs maximum flexure for slope 4H:1V.

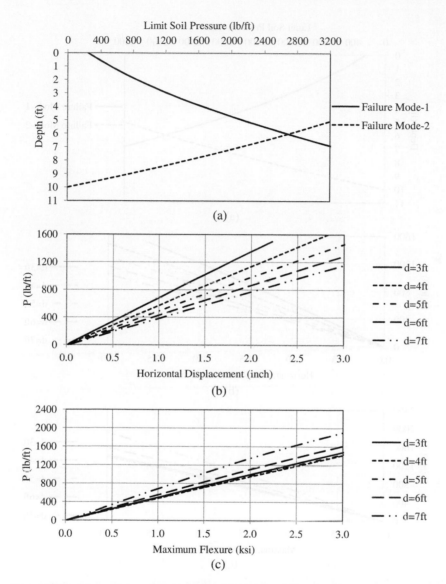

Figure A.33 Design chart for $c = 400$ psf and $\varphi = 0°$: a. limit soil curve; b. load vs horizontal displacement for slope 2H:1V; c. load vs maximum flexure for slope 2H:1V.

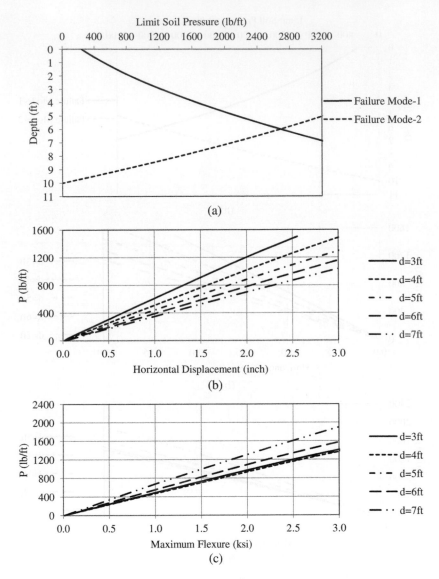

Figure A.34 Design chart for $c = 400$ psf and $\varphi = 0°$: a. limit soil curve; b. load vs horizontal displacement for slope 3H:1V; c. load vs maximum flexure for slope 3H:1V.

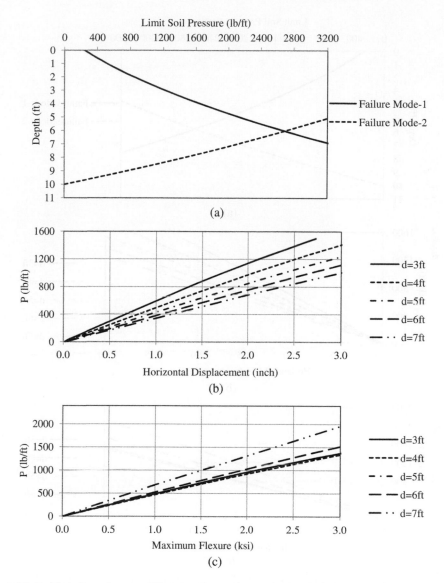

Figure A.35 Design chart for $c = 400$ psf and $\varphi = 0°$: a. limit soil curve; b. load vs horizontal displacement for slope 4H:1V; c. load vs maximum flexure for slope 4H:1V.

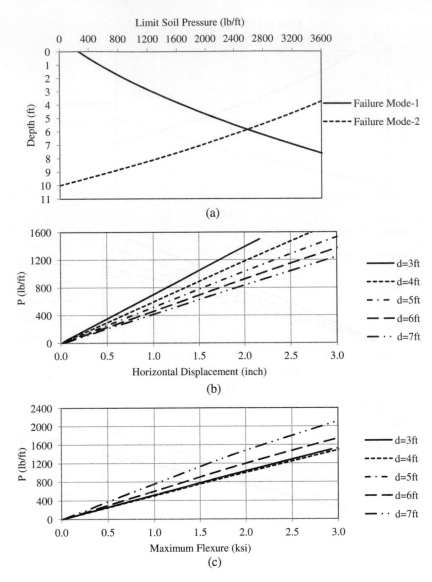

Figure A.36 Design chart for $c = 400$ psf and $\varphi = 10°$: a. limit soil curve; b. load vs horizontal displacement for slope 2H:1V; c. load vs maximum flexure for slope 2H:1V.

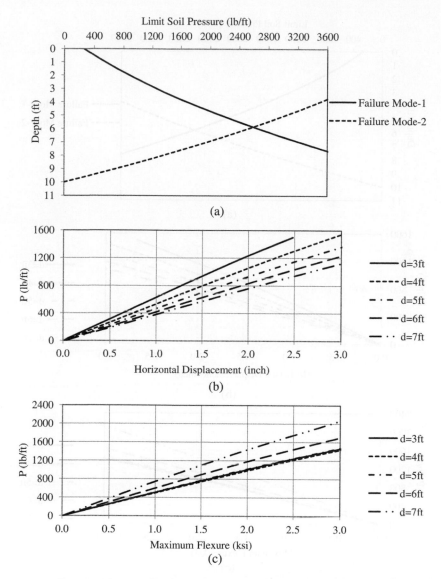

Figure A.37 Design chart for $c = 400$ psf and $\varphi = 10°$: a. limit soil curve; b. load vs horizontal displacement for slope 3H:1V; c. load vs maximum flexure for slope 3H:1V.

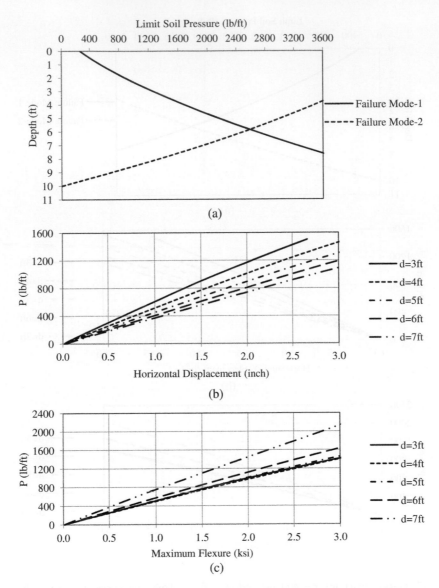

Figure A.38 Design chart for $c = 400$ psf and $\varphi = 10°$: a. limit soil curve; b. load vs horizontal displacement for slope 4H:1V; c. load vs maximum flexure for slope 4H:1V.

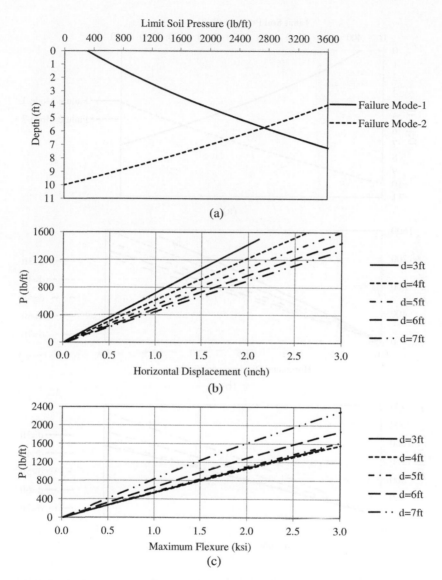

Figure A.39 Design chart for $c = 400$ psf and $\varphi = 20°$: a. limit soil curve; b. load vs horizontal displacement for slope 2H:1V; c. load vs maximum flexure for slope 2H:1V.

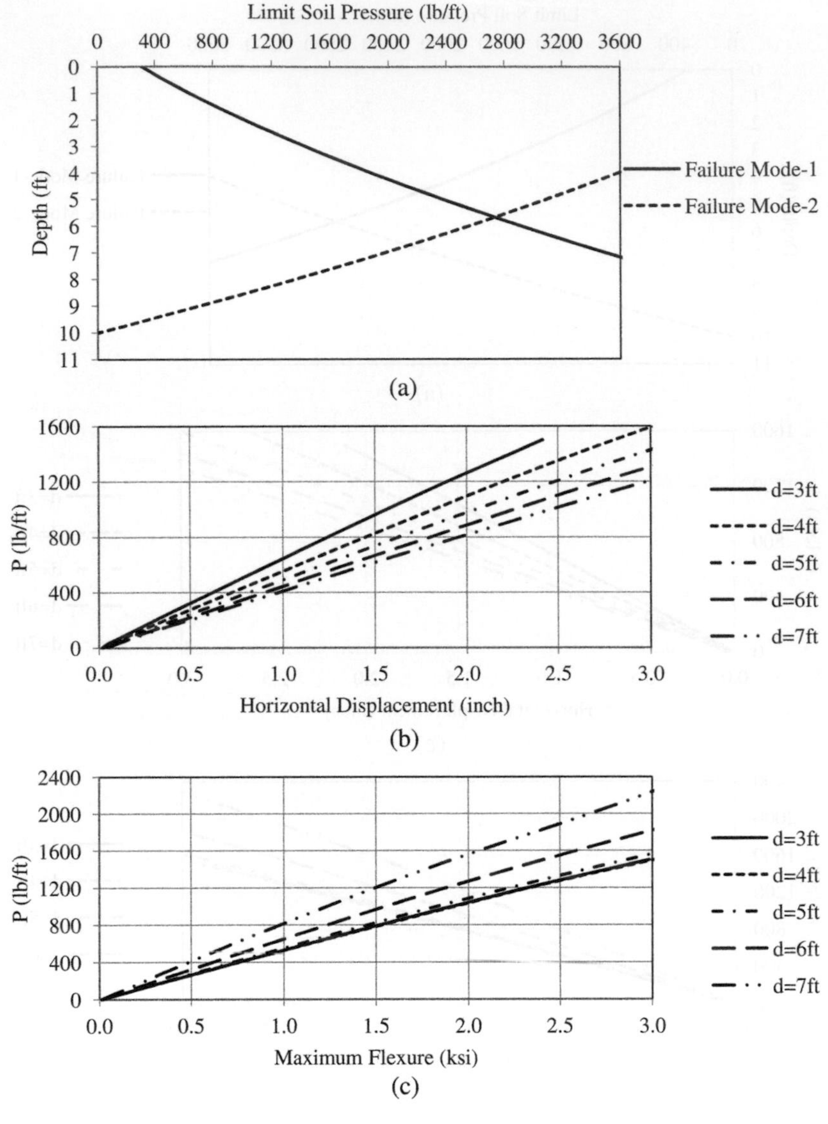

Figure A.40 Design chart for $c = 400$ psf and $\varphi = 20°$: a. limit soil curve; b. load vs horizontal displacement for slope 3H:1V; c. load vs maximum flexure for slope 3H:1V.

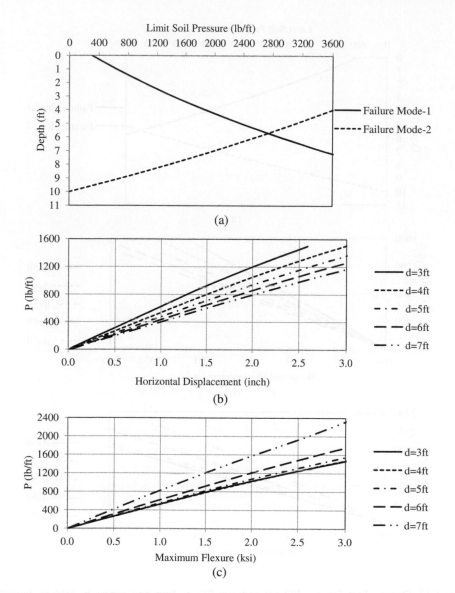

Figure A.41 Design chart for $c = 400$ psf and $\varphi = 20°$: a. limit soil curve; b. load vs horizontal displacement for slope 4H:1V; c. load vs maximum flexure for slope 4H:1V.

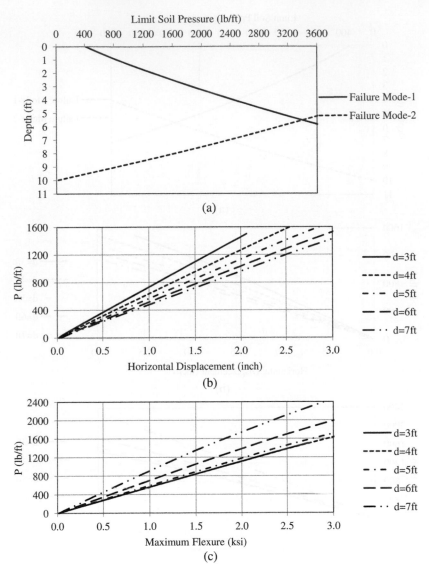

Figure A.42 Design chart for $c = 400$ psf and $\varphi = 30°$: a. limit soil curve; b. load vs horizontal displacement for slope 2H:1V; c. load vs maximum flexure for slope 2H:1V.

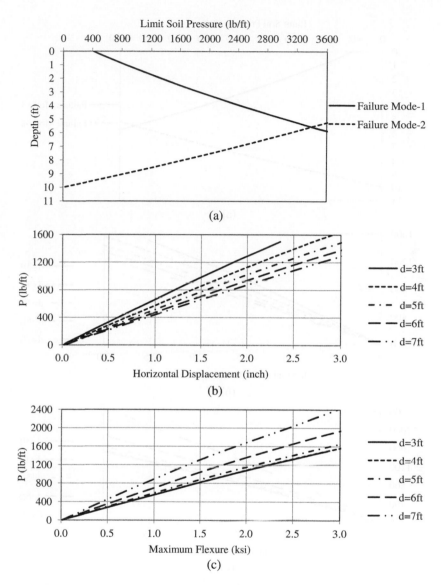

Figure A.43 Design chart for $c = 400$ psf and $\varphi = 30°$: a. limit soil curve; b. load vs horizontal displacement for slope 3H:1V; c. load vs maximum flexure for slope 3H:1V.

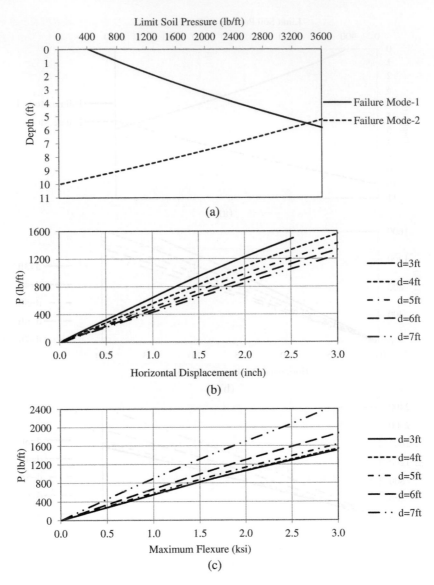

Figure A.44 Design chart for $c = 400$ psf and $\varphi = 30°$: a. limit soil curve; b. load vs horizontal displacement for slope 4H:1V; c. load vs maximum flexure for slope 4H:1V.

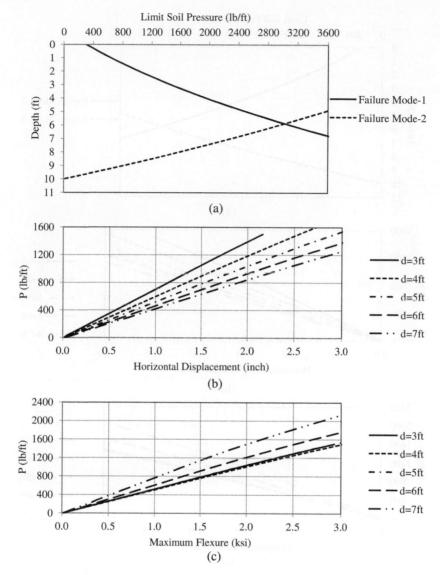

Figure A.45 Design chart for $c = 500$ psf and $\varphi = 0°$: a. limit soil curve; b. load vs horizontal displacement for slope 2H:1V; c. load vs maximum flexure for slope 2H:1V.

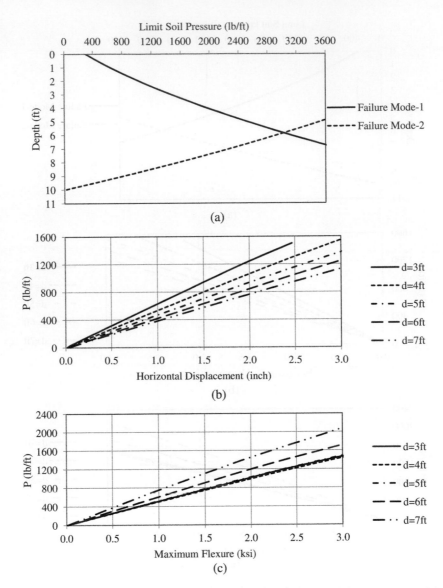

Figure A.46 Design chart for $c = 500$ psf and $\varphi = 0°$: a. limit soil curve; b. load vs horizontal displacement for slope 3H:1V; c. load vs maximum flexure for slope 3H:1V.

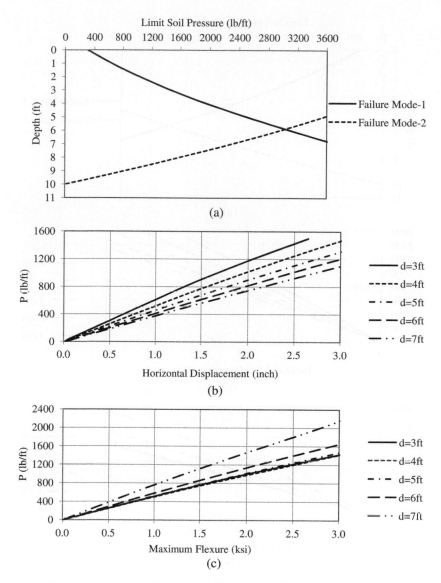

Figure A.47 Design chart for $c = 500$ psf and $\varphi = 0°$: a. limit soil curve; b. load vs horizontal displacement for slope 4H:1V; c. load vs maximum flexure for slope 4H:1V.

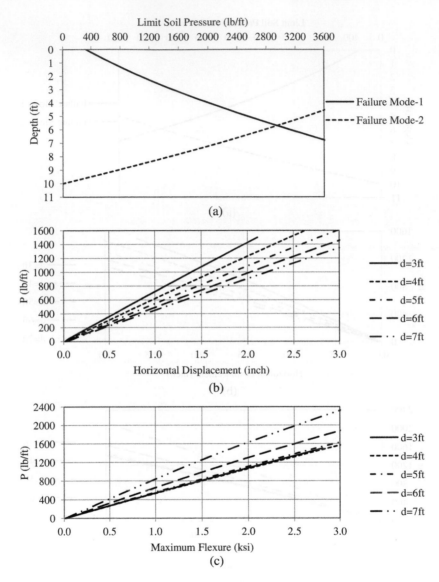

Figure A.48 Design chart for $c = 500$ psf and $\varphi = 10°$: a. limit soil curve; b. load vs horizontal displacement for slope 2H:1V; c. load vs maximum flexure for slope 2H:1V.

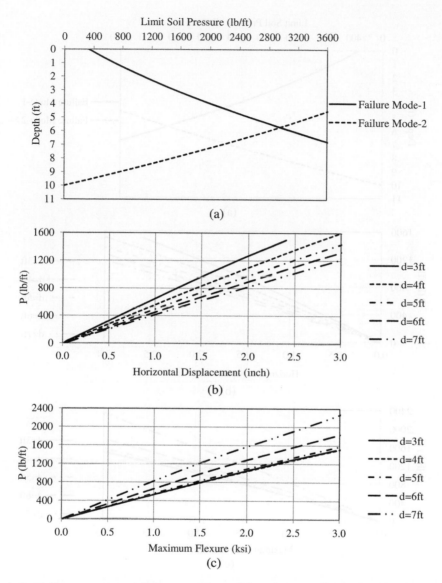

Figure A.49 Design chart for $c = 500$ psf and $\varphi = 10°$: a. limit soil curve; b. load vs horizontal displacement for slope 3H:1V; c. load vs maximum flexure for slope 3H:1V.

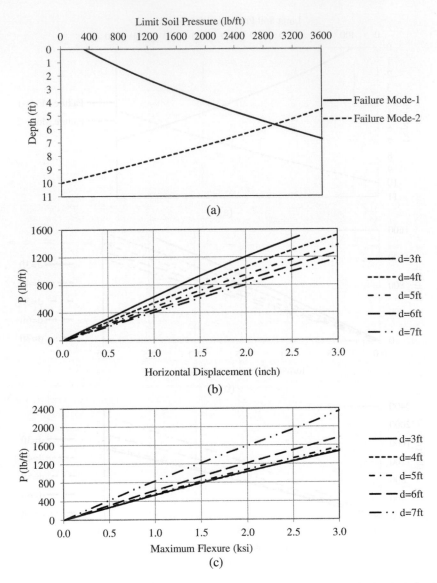

Figure A.50 Design chart for $c = 500$ psf and $\varphi = 10°$: a. limit soil curve; b. load vs horizontal displacement for slope 4H:1V; c. load vs maximum flexure for slope 4H:1V.

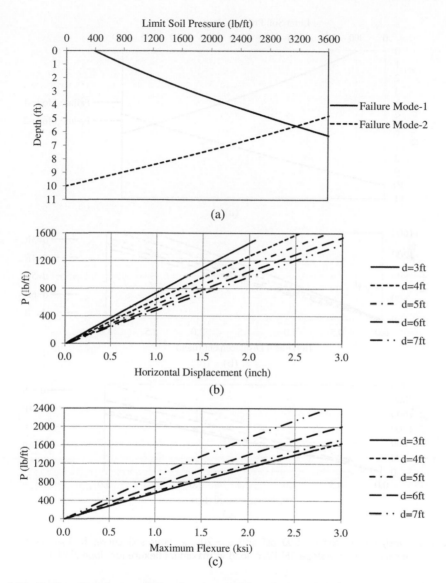

Figure A.5 I Design chart for $c = 500$ psf and $\varphi = 20°$: a. limit soil curve; b. load vs horizontal displacement for slope 2H:1V; c. load vs maximum flexure for slope 2H:1V.

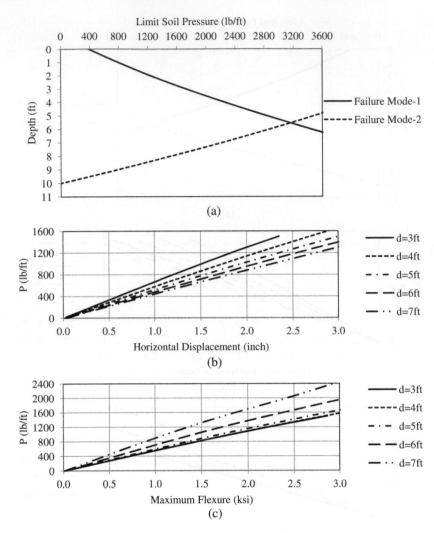

Figure A.52 Design chart for $c = 500$ psf and $\varphi = 20°$: a. limit soil curve; b. load vs horizontal displacement for slope 3H:1V; c. load vs maximum flexure for slope 3H:1V.

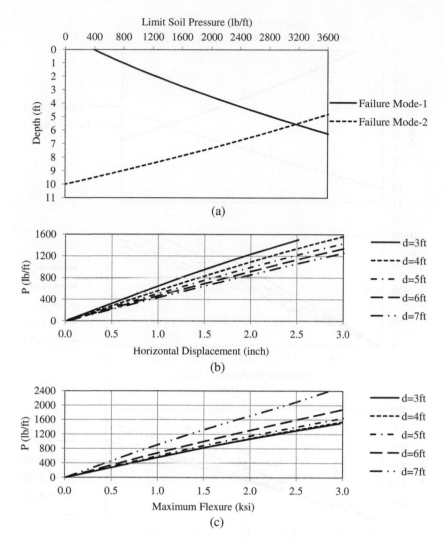

Figure A.53 Design chart for $c = 500$ psf and $\varphi = 20°$: a. limit soil curve; b. load vs horizontal displacement for slope 4H:1V; c. load vs maximum flexure for slope 4H:1V.

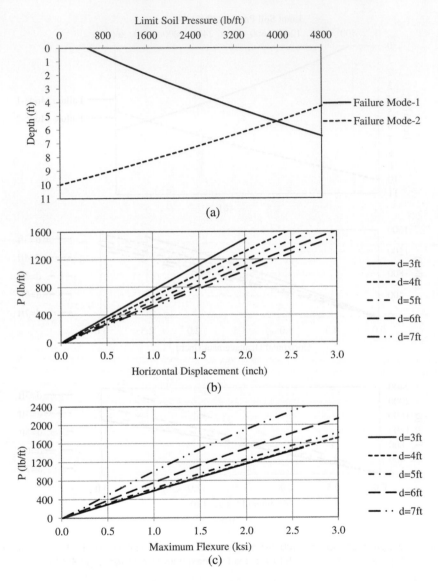

Figure A.54 Design chart for $c = 500$ psf and $\varphi = 30°$: a. limit soil curve; b. load vs horizontal displacement for slope 2H:1V; c. load vs maximum flexure for slope 2H:1V.

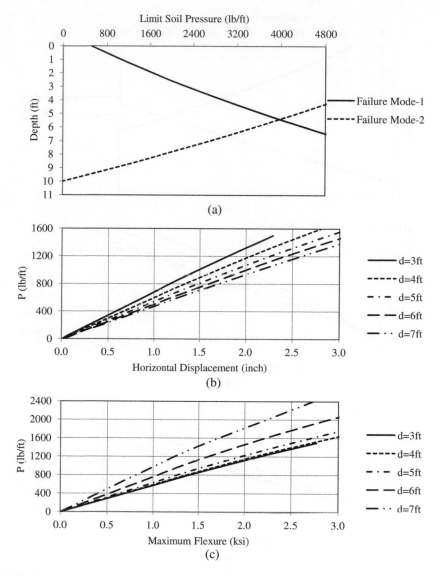

Figure A.55 Design chart for $c = 500$ psf and $\varphi = 30°$: a. limit soil curve; b. load vs horizontal displacement for slope 3H:1V; c. load vs maximum flexure for slope 3H:1V.

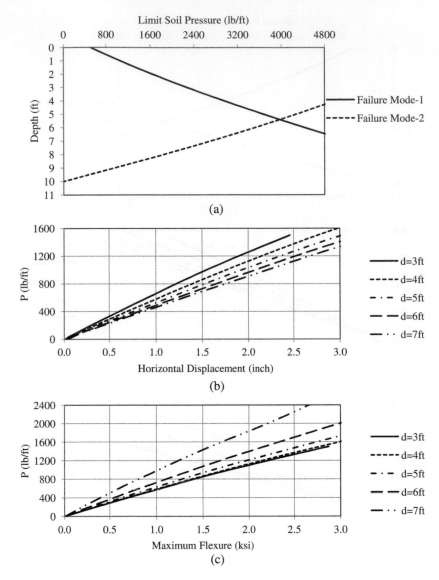

Figure A.56 Design chart for $c = 500$ psf and $\varphi = 30°$: a. limit soil curve; b. load vs horizontal displacement for slope 4H:1V; c. load vs maximum flexure for slope 4H:1V.

Appendix B

Sample calculations

APPROACH 1: SAMPLE CALCULATION USING ORDINARY METHOD OF SLICE

Problem definition

The current example presents the hand calculation of the design steps using the conventional method of slice approach (see Chapter 5, Section 5.7.1) for the US 287 slope (see Chapter 7, Section 7.1) depicted in Figure B.1 and having the soil properties presented in Table B.1.

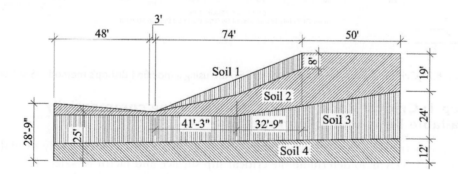

Figure B.1 Geometry of the US 287 slope.

Table B.1 Soil parameters of the US 287 slope.

Soil type	Friction angle φ (°)	Cohesion c (psf)	Unit weight γ (pcf)
1	10	130	125
2	23	130	125
3	15	250	130
4	35	3000	140

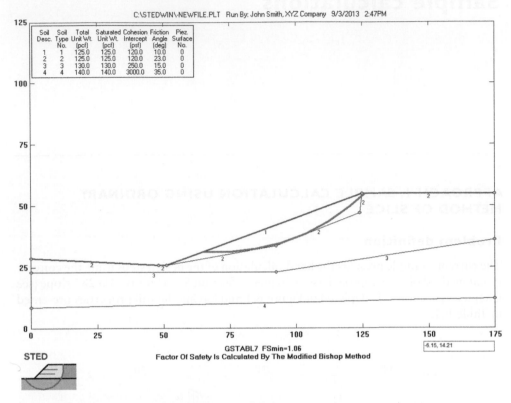

C:\STEDWIN\-NEWFILE.PLT   Run By: John Smith, XYZ Company   9/3/2013   2:47PM

Soil Desc.	Soil Type No.	Total Unit Wt. (pcf)	Saturated Unit Wt. (pcf)	Cohesion Intercept (psf)	Friction Angle (deg)	Piez. Surface No.
1	1	125.0	125.0	120.0	10.0	0
2	2	125.0	125.0	120.0	23.0	0
3	3	130.0	130.0	250.0	15.0	0
4	4	140.0	140.0	3000.0	35.0	0

GSTABL7 FSmin=1.06

-6.15, 14.21

STED

Factor Of Safety Is Calculated By The Modified Bishop Method

Figure B.2   Critical slip surface as modelled inGSTABL using a modified Bishop's method (FS = 1.06).

Step 1: Critical slope stability analysis using commercially available limit equilibrium method

The geometry with the soil parameters was modelled in the commercially available program GSTABL to determine the critical slip surface and factor of safety (FS). The critical slip surface derived is shown in Figure B.2. In addition, the details of the analysis results are included at the end of this section.

Step 2: Hand calculation of the unreinforced slope using the ordinary method of slice

Figure B.3 illustrates the slices in the critical slip surface used as the basis for calculation. Detail calculation of the unreinforced slope using ordinary method of slice is shown in Table B.2.

FS of unreinforced slope (ordinary method of slice):

$$FS = \frac{\sum c\beta + N \tan \varphi}{\sum W \sin \alpha}$$

$$FS = \frac{14282.75}{14106.99} = 1.012$$

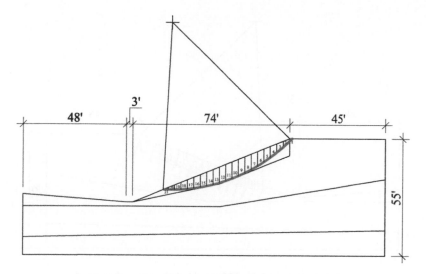

Figure B.3 Distribution of slices for hand calculation.

Table B.2 Detailed calculation of unreinforced slope using ordinary method of slice.

Slice no.	L1 (ft)	L2 (ft)	β (ft)	A (ft²)	γ (pcf)	W (lb)	α (°)	N (lb)	W sin α (lb)	c (psf)	φ (°)	c β + N tan φ (lb)
1	0.00	1.08	1.08	0.59	125.00	73.31	43.00	53.61	49.99	130.00	10.00	150.24
2	1.08	2.75	3.00	5.75	125.00	718.69	43.00	525.61	490.14	130.00	10.00	482.68
3	2.75	4.08	3.00	10.25	125.00	1281.19	41.00	966.92	840.53	130.00	10.00	560.49
4	4.08	5.17	3.00	13.88	125.00	1734.38	38.00	1366.71	1067.79	130.00	10.00	630.99
5	5.17	6.08	3.00	16.88	125.00	2109.38	35.00	1727.90	1209.89	130.00	10.00	694.68
6	6.08	6.75	3.00	19.25	125.00	2406.19	32.00	2040.56	1275.09	130.00	10.00	749.81
7	6.75	7.25	3.00	21.00	125.00	2625.00	30.00	2273.32	1312.50	130.00	10.00	790.85
8	7.25	7.52	3.00	22.15	125.00	2768.63	27.00	2466.86	1256.93	130.00	10.00	824.97
9	7.52	7.75	3.00	22.90	125.00	2862.50	25.00	2594.19	1209.69	130.00	10.00	847.43
10	7.75	7.83	3.00	23.37	125.00	2921.25	22.00	2708.54	1094.32	130.00	10.00	867.59
11	7.83	7.67	3.00	23.25	125.00	2906.25	20.00	2730.98	994.00	130.00	10.00	871.55
12	7.67	7.42	3.00	22.63	125.00	2828.81	18.00	2690.36	874.15	130.00	10.00	864.38
13	7.42	7.08	3.00	21.75	125.00	2718.75	15.00	2626.11	703.66	130.00	10.00	853.05
14	7.08	6.50	3.00	20.37	125.00	2546.81	13.00	2481.54	572.91	130.00	10.00	827.56
15	6.50	5.92	3.00	18.63	125.00	2328.13	11.00	2285.36	444.23	130.00	10.00	792.97
16	5.92	5.17	3.00	16.63	125.00	2078.19	10.00	2046.62	360.87	130.00	10.00	750.87
17	5.17	4.25	3.00	14.13	125.00	1765.69	7.00	1752.53	215.18	130.00	10.00	699.02
18	4.25	3.33	3.00	11.37	125.00	1421.25	4.00	1417.79	99.14	130.00	10.00	639.99
19	3.33	2.17	3.00	8.25	125.00	1030.69	2.00	1030.06	35.97	130.00	10.00	571.63
20	2.17	1.00	3.00	4.75	125.00	593.81	0.00	593.81	0.00	130.00	10.00	494.71
21	1.00	0.00	2.25	1.13	125.00	140.63	0.00	140.63	0.00	130.00	10.00	317.30
Total									14106.99		Total	14282.75

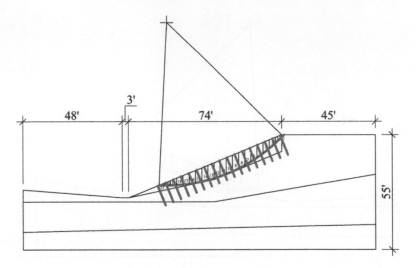

Figure B.4 Location of the RPPs in relation to the different slices.

Step 3: Design of reinforced slope

Step 3A: Selection of RPP

Based on different commercially available samples, the following RPP properties were selected for the design:

- Dimension of RPPs: 3.5 inch × 3.5 inch
- Ultimate flexural strength: 4 ksi
- Allowable flexural strength: $4 \times 0.35 = 1.4$ ksi

Step 3B: Hand calculation of the reinforced slope using the ordinary method of slice

To determine the FS of the reinforced slope, it is important to select a trial RPP spacing for the analysis. The current example assumed an RPP spacing of 4 ft centre-to-centre (c/c) as presented in Figure B.4. The final spacing of the RPPs should be selected from a few iterations of different RPP spacing, based on the target FS.

The current analysis of slope reinforcement was performed considering the specific limit resistance of the RPPs based on the depth of the slip surface. The limit resistance of the RPPs was determined based on the limit resistance curve presented in Table B.3.

FS of reinforced slope (ordinary method of slice):

$$FS = \frac{\sum c\beta + N \tan \varphi + P}{\sum W \sin \alpha}$$

$$FS = \frac{20444.75}{14106.99} = 1.45$$

A similar calculation can be repeated with a smaller RPP spacing (<4 ft c/c) to increase the FS.

Table B.3 Detailed calculation of reinforced slope using ordinary method of slice.

Slice no.	α (°)	W (lb)	N (lb)	$W \sin \alpha$ (lb)	c (psf)	φ (°)	$c\beta + N \tan \varphi$ (lb)	d (ft)	P_1 (Figure B.5) (lb)	P_2 (Figure B.6) (lb)	P (lb) (avg)	$c\beta + N \tan \varphi + P$ (lb)
1	43	73.3	53.6	50.0	130.0	10.0	150.2	0.0	0.0	0.0	0.0	150.2
2	43	718.7	525.6	490.1	130.0	10.0	482.7	3.0	550.0	600.0	565.0	1047.7
3	41	1281.2	966.9	840.5	130.0	10.0	560.5	3.0	550.0	600.0	565.0	1125.5
4	38	1734.4	1366.7	1067.8	130.0	10.0	631.0	4.0	430.0	520.0	457.0	1088.0
5	35	2109.4	1727.9	1209.9	130.0	10.0	694.7	5.0	350.0	410.0	368.0	1062.7
6	32	2406.2	2040.6	1275.1	130.0	10.0	749.8	5.0	350.0	410.0	368.0	1117.8
7	30	2625.0	2273.3	1312.5	130.0	10.0	790.8	6.0	210.0	350.0	252.0	1042.8
8	27	2768.6	2466.9	1256.9	130.0	10.0	825.0	7.0	200.0	300.0	230.0	1055.0
9	25	2862.4	2594.2	1209.7	130.0	10.0	847.4	7.0	200.0	300.0	230.0	1077.4
10	22	2921.3	2708.5	1094.3	130.0	10.0	867.6	7.0	200.0	300.0	230.0	1097.6
11	20	2906.3	2731.0	994.0	130.0	10.0	871.5	7.0	200.0	300.0	230.0	1101.5
12	18	2828.8	2690.4	874.2	130.0	10.0	864.4	7.0	200.0	300.0	230.0	1094.4
13	15	2718.8	2626.1	703.7	130.0	10.0	853.1	7.0	200.0	300.0	230.0	1083.1
14	13	2546.4	2481.5	572.9	130.0	10.0	827.6	0.0	0.0	0.0	0.0	827.6
15	11	2328.1	2285.4	444.2	130.0	10.0	793.0	6.0	210.0	350.0	252.0	1045.0
16	10	2078.2	2046.6	360.9	130.0	10.0	750.9	5.0	350.0	410.0	368.0	1118.9
17	7	1765.7	1752.5	215.2	130.0	10.0	699.0	4.0	430.0	520.0	457.0	1156.0
18	4	1421.3	1417.8	99.1	130.0	10.0	640.0	0.0	0.0	0.0	0.0	640.0
19	2	1030.7	1030.1	36.0	130.0	10.0	571.6	3.0	550.0	600.0	565.0	1136.6
20	0	593.8	593.8	0.0	130.0	10.0	494.7	3.0	550.0	600.0	565.0	1059.7
21	0	140.6	140.6	0.0	130.0	10.0	317.3	0.0	0.0	0.0	0.0	317.3
Total				14106.99							Total	20444.75

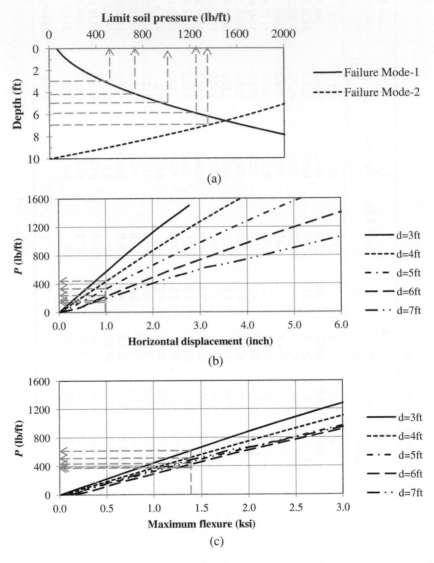

Figure B.5 Limit soil pressure design charts for $c = 100$ psf and $\varphi = 10°$: a. limit soil curve; b. load vs horizontal displacement for slope 3H:1V; c. load vs maximum flexure for slope 3H:1V (adapted from Appendix A, Figure A.1 and based on Ito and Matsui (1975)).

Output of GSTABL program, used to determine the critical slip surface in Step 1 above

```
*** GSTABL7 ***

** GSTABL7 by Garry H. Gregory, P.E. **
```

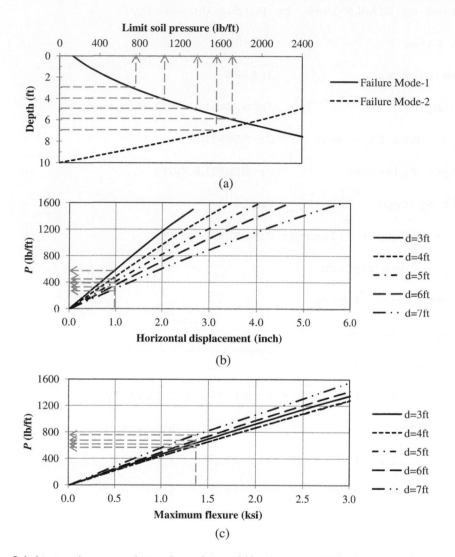

Figure B.6 Limit soil pressure design charts for $c = 200$ psf and $\varphi = 10°$: a. limit soil curve; b. load vs horizontal displacement for slope 3H:1V; c. load vs maximum flexure for slope 3H:1V (adapted from Appendix A, Figure A.13, and based on Ito and Matsui (1975)).

** Version 1.0, January 1996; Version 1.16, May 2000 **

--Slope Stability Analysis--

Simplified Janbu, Modified Bishop

or Spencer's Method of Slices

(Based on STABL6-1986, by Purdue University)

Run Date: 9/3/2013

Time of Run: 2:47PM

Run By: John Smith, XYZ Company

Input Data Filename: C:-NEWFILE.

Output Filename: C:-NEWFILE.OUT

Unit System: English

Plotted Output Filename: C:-NEWFILE.PLT

problem description

boundary coordinates

4 Top Boundaries

10 Total Boundaries

Boundary X-Left Y-Left X-Right Y-Right Soil Type

No.	(ft)	(ft)	(ft)	(ft)	Below Bnd
1	0.00	28.75	48.00	25.75	2
2	48.00	25.75	51.00	25.75	2
3	51.00	25.75	125.00	55.00	1
4	125.00	55.00	175.00	55.00	2
5	51.00	25.75	92.25	33.75	2
6	92.25	33.75	124.00	47.00	2
7	124.00	47.00	125.00	55.00	2
8	0.00	23.00	92.25	23.00	3
9	92.25	23.00	175.00	36.00	3
10	0.00	8.50	175.00	12.00	4

isotropic soil parameters

4 Type(s) of Soil

Soil Type No.	Total Unit Wt. (pcf)	Saturated Unit Wt. (pcf)	Cohesion Intercept (psf)	Friction Angle (deg)	Pore Pressure Param.	Pressure Constant (psf)	Piez. Surface No.
1	125.0	125.0	120.0	10.0	0.00	0.0	0
2	125.0	125.0	120.0	23.0	0.00	0.0	0
3	130.0	130.0	250.0	15.0	0.00	0.0	0
4	140.0	140.0	3000.0	35.0	0.00	0.0	0

Trial Failure Surface Specified By 8 Coordinate Points

Point No.	X-Surf (ft)	Y-Surf (ft)
1	64.49	31.08
2	74.49	31.09
3	84.41	32.36
4	94.09	34.88
5	103.37	38.60
6	112.11	43.46
7	120.16	49.39
8	126.04	55.00

Circle Center At X = 69.5 ; Y = 109.4 and Radius, 78.5

* * Factor Of Safety Is calculated by the modified bishop method * *

Factor Of Safety for the preceding specified surface = 1.064

Table 1 - Individual data on the 9 Slices

Water Water Tie Tie Earthquake

Force Force Force Force Force Surcharge

Slice Width Weight Top Bot Norm Tan Hor Ver Load

No.	(ft)	(lbs)	(lbs)	(lbs)	(lbs)	(lbs)	(lbs)	(lbs)	(lbs)
1	10.0	2465.6	0.0	0.0	0.0	0.0	0.0	0.0	0.0
2	9.9	6535.3	0.0	0.0	0.0	0.0	0.0	0.0	0.0
3	9.7	8771.4	0.0	0.0	0.0	0.0	0.0	0.0	0.0
4	9.3	9136.5	0.0	0.0	0.0	0.0	0.0	0.0	0.0
5	8.7	7808.8	0.0	0.0	0.0	0.0	0.0	0.0	0.0
6	8.0	5102.6	0.0	0.0	0.0	0.0	0.0	0.0	0.0
7	4.7	1400.3	0.0	0.0	0.0	0.0	0.0	0.0	0.0
8	0.1	18.2	0.0	0.0	0.0	0.0	0.0	0.0	0.0
9	1.0	64.5	0.0	0.0	0.0	0.0	0.0	0.0	0.0

Table 2 - Base stress data on the 9 Slices

Slice Alpha X-Coord. Base Available Mobilized

No. (deg)	Slice Cntr	Leng.	Shear Strength	Shear Stress
*	(ft)	(ft)	(psf)	(psf)
1 0.04	69.49	10.00	163.45	0.19
2 7.30	79.45	10.00	231.26	82.98
3 14.59	89.25	10.00	268.21	220.92
4 21.84	98.73	10.00	275.31	340.02
5 29.08	107.74	10.00	254.12	379.48
6 36.38	116.13	10.00	206.55	302.68

7   43.65   122.51   6.49   148.99   148.83

8   43.65   124.93   0.19   126.57   64.40

9   43.65   125.52   1.44   105.99   30.97

Sum of the resisting forces (including pier/pile,

tieback, and reinforcing forces if applicable) = 15137.83 (lbs)

Average available shear strength (including tieback,

pier/pile, and reinforcing forces if applicable) = 222.20(psf)

Sum of the driving forces = 14286.11 (lbs)

Average mobilized shear stress = 209.70(psf)

Total length of the failure surface = 68.13(ft)

APPROACH 2: SAMPLE CALCULATION USING INFINITE SLOPE METHOD

Problem definition

The current example presents the hand calculation of the design steps using the infinite slope method (see Chapter 5, Section 5.7.2) for the US 287 slope (see Chapter 7, Section 7.1) depicted in Figure B.7 and having the soil properties presented in Table B.4.

Step 1: Determine FS of the unreinforced slope

- $d = 7\,\text{ft}$
- $L = 126.5\,\text{ft}$
- $\beta = 18.42°$
- $\gamma' = 125\,\text{pcf}$

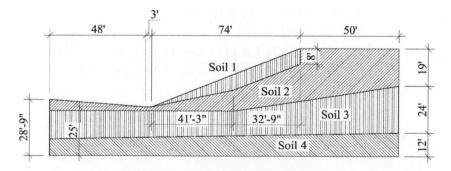

Figure B.7 Geometry of the US 287 slope.

Table B.4 Soil parameters of the US 287 slope.

Soil type	Friction angle φ (°)	Cohesion c (psf)	Unit weight γ (pcf)
1	10	130	125
2	23	130	125
3	15	250	130
4	35	3000	140

- $\gamma_{sat} = 125\,\text{pcf}$
- $c = 130\,\text{psf}$
- $\phi = 10°$

FS of unreinforced slope (infinite slope method):

$$h = d/\cos\beta = 7/\cos 18.25 = 7.37\,\text{ft}$$

$$FS = \frac{c' + h\gamma'\cos^2\beta * \tan\varphi'}{\gamma_{sat} * h * \sin\beta * \cos\beta}$$

$$FS = \frac{130 + 7.37 * 125 * \cos^2 18.42 * \tan 10}{125 * 7.37 * \sin 18.42 * \cos 18.42}$$

$$FS = 1.0$$

Step 2: Using the depth of failure, determine the limit resistance, P, from the appropriate design chart in Appendix A (examples shown in Figures B.8 and B.9). The limit resistance from Figure B.8 and B.9 are summarized in Table B.5.

Step 3: Select an RPP spacing and determine the FS of the reinforced slope

FS of reinforced slope (infinite slope method):

$$P_{design} = 230\,\text{lb/ft}$$

$$s = 3\,\text{ft c/c}$$

$$FS = \frac{c'L + hL\gamma'\cos^2\beta \times \tan\varphi' + \left(\frac{L}{S} + 1\right) \times P}{\gamma_{sat} \times hL \times \sin\beta \times \cos\beta}$$

$$FS = \frac{130 \times 126.5 \times 7.37 \times 126.5 \times 125 \times \cos^2 18.42 \times \tan 10 + \left(\frac{126.5}{3} + 1\right) \times 230}{125 \times 7.37 \times 126.5 \times \sin 18.42 \times \cos 18.42}$$

$$FS = 1.26$$

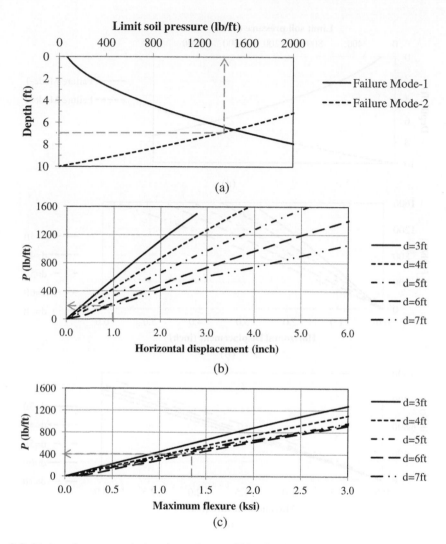

Figure B.8 Limit soil pressure design charts for $c = 100\,psf$ and $\varphi = 10°$: a. Limit soil curve; b. load vs horizontal displacement for slope 3H:1V; c. load vs maximum flexure for slope 3H:1V (adapted from Appendix A, Figure A.1, and based on Ito and Matsui (1975)).

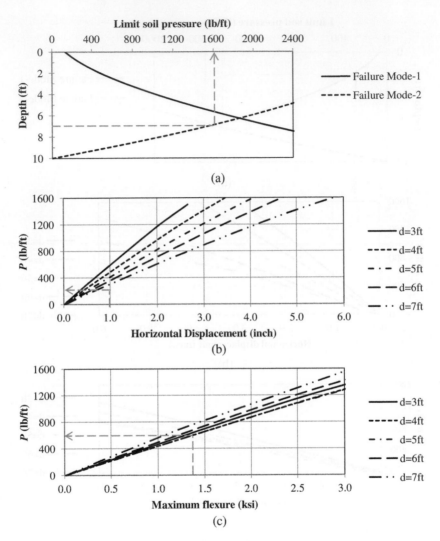

Figure B.9 Limit soil pressure design charts for $c = 200$ psf and $\varphi = 10°$: a. limit soil curve; b. load vs horizontal displacement for slope 3H:1V; c. load vs maximum flexure for slope 3H:1V (adapted from Appendix A, Figure A.13, and based on Ito and Matsui (1975)).

Table B.5 Summary of limit resistance.

c (psf)	φ (°)	d (ft)	P_1 (Figure B.8) (lb)	P_2 (Figure B.9) (lb)	P_{design} (lb) (avg)
130.0	10.0	7.0	200.0	300.0	230.0

Sample excel sheet

Design of slope stabilisation using recycled plastic pins	
Method of analysis: infinite slope analysis	
Design Parameters	
Length of RPP (ft)	10
Cohesion, c (psf)	130
Friction angle, φ (°)	10
Bulk unit weight of failure zone, γ' (pcf)	125
Saturated unit weight of failure zone (pcf)	125
Depth of failed zone, d (ft)	7
Slope ratio: vertical face	1
Slope ratio: horizontal face	3
Slope ratio (ver/hor)	1/3
Height of slope (ft)	20
Length of the slope face (ft)	60
1/3 length near crest (ft)	20.00
Spacing of RPPs at top 1/3 (ft)	3
2/3 length near toe (ft)	40.00
Spacing of RPPs at bottom 2/3 (ft)	3
Load resistance from RPPs, P (lb/ft)	230
Number of RPPs	20
Calculation	
Slope angle, β (°)	18.42
Length of slope face, L (ft)	63.29
Vertical depth of failure zone (ft)	7.38
Results	
Initial FS (no reinforcement)	1.00
Final FS (with RPPs)	1.26

*Excel Sheet is included with the book.

References

Abramson, L., Lee, T., Sharma, S. & Boyce, G. (2002) *Slope Stability and Stabilization Methods*. 2nd edition. New York, NY, John Wiley & Sons.

Ahmed, F.S. (2013) *Engineering Characteristics of Recycled Plastic Pin, Lumber and Bamboo for Soil Slope Stabilization*. MS Thesis. University of Texas at Arlington.

American Society for Testing Materials. (2009a) *D6108-09. Standard Test Method for Compressive Properties of Plastic Lumber and Shapes*. Section 8, Vol. 08.03.

American Society for Testing Materials. (2009b) *D6109-09. Standard Test Methods for Flexural Properties of Unreinforced and Reinforced Plastic Lumber and Related Products*. Section 8, Vol. 08.03.

American Society for Testing Materials. (2016) *D2344-16. Standard Test Method for Short-Beam Strength of Polymer Matrix Composite Materials and Their Laminates*. Section 15, Vol. 15.03.

Andrady, A.L. & Neal, M.A. (2009) Applications and societal benefits of plastics. *Philosophical Transactions of the Royal Society B: Biological Sciences*, 364 (1526), 1977–1984.

Aubeny, C.P. & Lytton, R.L. (2004) Shallow slides in compacted high plasticity clay slopes. *Journal of Geotechnical and Geoenvironmental Engineering*, 130 (7), 717–727.

Bassi, A. (2017) Biotechnology for the management of plastic wastes. In: *Current Developments in Biotechnology and Bioengineering: Solid Waste Management*. Amsterdam, Elsevier. pp. 293–310.

Berg, R.R., Christopher, B.R. & Samtani, N.C. (2009) *Design of Mechanically Stabilized Earth Walls and Reinforced Soil Slopes–Volume II (No. FHWA-NHI-10-025)*.

Bowders, J.J., Loehr, J.E., Salim, H. & Chen, C.W. (2003) Engineering properties of recycled plastic pins for slope stabilization. *Transportation Research Record: Journal of the Transportation Research Board*, 1849 (1), 39–46.

Breslin, V.T., Senturk, U. & Berndt, C.C. (1998) Long-term engineering properties of recycled plastic lumber in pier construction. *Resources, Conservation and Recycling*, 23, 243–258.

Cai, F. & Ugai, K. (2003) Reinforcing mechanism of anchors in slopes: A numerical comparison of results of LEM and FEM. *International Journal for Numerical and Analytical Methods in Geomechanics*, 27, 549–564.

Chen, C.W., Salim, H., Bowders, J., Loehr, E. & Owen, J. (2007) Creep behavior of recycled plastic lumber in slope stabilization applications. *Journal of Materials in Civil Engineering*, 19 (2), 130–138.

Chen, L. & Poulos, H.G. (1993) Analysis of pile-soil interaction under lateral loading using infinite and finite elements. *Computers and Geotechnics*, 15 (4), 189–220.

Chen, L.T. and Poulos, H.G. (1997) Piles subjected to lateral soil movements. *Journal of Geotechnical and Geoenvironmental Engineering*, 123 (9), 802–811.

Day, R.W. (1996) Design and repair for surficial slope failures. *Practice Periodical on Structural Design and Construction*, 1 (3), 83–87.

Day, R.W. & Axten, G.W. (1989) Surficial stability of compacted clay slopes. *Journal of Geotechnical Engineering*, 115 (4), 577–580.

Department of the Army USA, Technical Manual TM 5-818-7, Foundations in Expansive Soils, 1 September 1983.

Elias, V., Christopher, B. & Berg, R. (2001) *Mechanically Stabilized Earth Walls and Reinforced Soil Slopes: Design and Construction Guidelines*. Washington, DC, Report FHWA-NHI-00-043, Federal Highway Administration.

Evans, D.A. (1972) *Slope Stability Report*. Los Angeles, CA, Slope Stability Committee, Department of Building and Safety.

Fay, L., Akin, M. & Shi, X. (2012) *Cost-Effective and Sustainable Road Slope Stabilization and Erosion Control*, Vol. 430. Washington, DC, Transportation Research Board.

Fredlund, D.G., Rahardjo, H., & Fredlund, M.D. (2012) Unsaturated soil mechanics in engineering practice. John Wiley & Sons.

Fredlund, D.G., Morgenstern, N.R. & Widger, R.A. (1978) The shear strength of unsaturated soils. *Canadian Geotechnical Journal*, 15 (3), 313–321.

Gopu, V.K.A. & Seals, R.K. (1999) Mechanical properties of recycled plastic lumber and implications in structural design. In: *Proceedings of the International Composites Expo '99, 10–12 May 1999, Cincinnati, OH*. Boca Raton, FL, CRC Press.

Gray, D.H. & Sotir, R.B. (1996) *Biotechnical and Soil Bioengineering Slope Stabilization: A Practical Guide for Erosion Control*. New York, NY, John Wiley & Sons.

Griffiths, D.V. & Lane, P.A. (1999) Slope stability analysis by finite elements. *Geotechnique*, 49 (3), 387–403.

Hoornweg, D. & Bhada-Tata, P. (2012) *What a Waste: A Global Review of Solid Waste Management. Urban Development Series; Knowledge Papers No. 15*. Washington, DC, World Bank.

Hossain, J. (2012) *Geohazard Potential of Rainfall Induced Slope Failure on Expansive Clay*. PhD Dissertation. University of Texas at Arlington.

Hossain, J., Khan, M.S., Hossain, M.S. & Ahmed, A. (2016) Determination of active zone in expansive clay in North Texas through field instrumentation. In: *Proc. 95th Annual Meeting of Transportation Research Board, 10–14 January 2016, Washington, DC*.

Hossain, M.S., Maganti, D. & Hossain, J. (2010) Assessment of geo-hazard potential and site investigations using resistivity imaging. *International Journal of Environmental Technology and Management*, 13 (2), 116–129.

Inci, G. (2008) *Numerical Modeling of Desiccation Cracking in Compacted Soils*. Goa, International Association for Computer Methods and Advances in Geomechanics (IACMAG).

Ito, T. & Matsui, T. (1975) Methods to estimate lateral force acting on stabilizing piles. *Soils and Foundations*, 15 (4), 43–59.

Jutkofsky, W.S., Sung, J.T. & Negussey, D. (2000) Stabilization of embankment slope with geofoam. *Transportation Research Record: Journal of the Transportation Research Board*, 1736 (1), 94–102.

Kandaris, P.M. (2007) Use of gabions for localized slope stabilization in difficult terrain. In: *Proceedings of the 37th US Symposium on Rock Mechanics, Vail, CO, 7–9 June 2007*.

Kayyal, M.K. & Wright S.G. (1991) *Investigation of Long-Term Strength Properties of Paris and Beaumont Clays in Earth Embankments. Research Report 1195-2F*. Center for Transportation Research, University of Texas at Austin.

Khan, M.S. & Hossain, M.D. (2015) Effects of shrinkage and swelling behavior of high plastic clay on the performance of a highway slope reinforced with recycled plastic pin. In: *Transportation Research Board 94th Annual Meeting (No. 15-5423)*.

Khan, M.S., Kibria, G., Hossain, M.S., Hossain, J. & Lozano, N. (2013) Performance evaluation of a slope reinforced with recycled plastic pin. In: *Proceedings of the 2013 Geo-Congress,*

San Diego, CA, 3–7 March 2013. Geotechnical Special Publication No. 231. American Society of Civil Engineers. pp. 1733–1742.

Khan, M.S., Hossain, S. & Kibria, G. (2015). Slope Stabilization Using Recycled Plastic Pins. *Journal of Performance of Constructed Facilities*, 30(3), 04015054.

Krishnaswamy, P. & Francini, R. (2000). Long-Term Durability of RPL in Structures. *R-2000 5th World Congress and Envirotech Trade Show*, Toronto, Canada.

Lampo, R. & Nosker, T.J. (1997) *Development and Testing of Plastic Lumber Materials for Construction Applications, USACERL Technical Report 97/95*. Champaign, IL, US Army Corps of Engineers, Construction Engineering Research Laboratories.

Lampo, R., Nosker, T., Barno, D., Busel, J., Mäher, A., Dutta, P. & Odello, R. (1998) *Development and Demonstration of FRP Composite Fender and Sheet Piling Systems, USACERL Technical Report 98/121*. Champaign, IL, US Army Corps of Engineers, Construction Engineering Research Laboratories. pp. 20–21.

Loehr, J.E. & Bowders, J.J. (2007) *Slope Stabilization Using Recycled Plastic Pins – Phase III, Final Report: RI98-007D*. Jefferson City, Missouri Department of Transportation.

Loehr, J.E., Bowders, J., Owen, J., Sommers, L. & Liew, L. (2000) Stabilization of slopes using recycled plastic pins. *Journal of the Transportation Research Board*, 1714, 1–8.

Loehr, J.E., Bowders, J.J., Chen, C.W., Chandler, K.S., Carr, P.H. & Fennessey, T.W. (2004) Slope stabilization using recycled plastic pins. In: Henthorne, R. (ed.) *Proceedings of the 55th Highway Geology Symposium.*

Loehr, J.E., Fennessey, T.W. & Bowders, J.J. (2007) Stabilization of surficial slides using recycled plastic reinforcement. *Transportation Research Record: Journal of the Transportation Research Board*, 1989 (1), 79–87.

Lynch, J.K., Nosker, T.J., Renfree, R.W., Krishnaswamy, P. & Francini, R. (2001) Weathering effects on mechanical properties of recycled HDPE-based plastic lumber. In: *Proceedings, ANTEC Conference, 6–10 May 2001, Dallas, TX*. Society of Plastics Engineers.

McCormick, W. & Short, R. (2006) Cost effective stabilization of clay slopes and failures using plate piles. In: *Proc. IAEG2006*. London, The Geological Society of London. pp. 1–7.

McLaren, M.G. (1995) Recycled plastic lumber & shapes design and specifications. In: *Restructuring: America and Beyond*. American Society of Civil Engineers. pp. 819–833.

McLaren, M.G. & Pensiero, J.P. (1999) Simplified design of recycled plastic as structural materials. In: *Proceedings of the International Composites Expo '99, 10–12 May 1999, Cincinnati, OH*. Boca Raton, FL, CRC Press.

Malcolm, G.M. (1995) Recycled plastic lumber and shapes design and specifications. In: *Proc. Structures Congress 13, 2–5 April 1995, Boston, MA*.

Nelson, J. & Miller, J.D. (1992) *Expansive Soils: Problems and Practice in Foundation and Pavement Engineering*. New York, NY, John Wiley & Sons.

Nosker, T. (1989) Improvements in the properties of commingled waste by the selective mixing of plastics waste. In: *Proc. SPE Recycling RETEC, Charlotte, NC*.

Nosker, T. (1999) The development of polyolefin based oriented glass fiber building materials. In: *Proc. ANTEC*. Society of Plastics Engineers.

Nosker, T. & Renfree, R. (2000) Recycled plastic lumber: From park benches to bridges. In: *Approved for Proceedings of R'2000 5th World Congress, Toronto, Canada*.

Parra, J.R., Loehr, J.E., Hagemeyer, D.J. & Bowders, J.J. (2003) Field performance of embankments stabilized with recycled plastic reinforcement. *Transportation Research Record: Journal of the Transportation Research Board*, 1849 (1), 31–38.

Parra, J., Ang, E. & Loehr, E. (2004) Sources of uncertainty in lateral resistance of slender reinforcement used for slope stabilization. In: *Proceedings of the 2004 GeoSupport Conference, Orlando, FL, 29–31 December 2004*. Geotechnical Special Publication No. 124. American Society of Civil Engineers. pp. 187–198.

Pearlman, S.L., Campbell, B.D. & Withiam, J.L. (1992) Slope stabilization using in-situ earth reinforcements. In: *Proceedings of the Conference on Stability and Performance of Slopes and Embankments II*. Geotechnical Special Publication No. 31. American Society of Civil Engineers. pp. 1333–1348.

PlasticsEurope (2008) *The Compelling Facts About Plastics 2007: An Analysis of Plastics Production, Demand and Recovery for 2007 in Europe*. Brussels, PlasticsEurope.

PlasticsEurope (2013) *Plastics – The Facts 2013: An Analysis of European Latest Plastics Production, Demand and Waste Data*. Brussels, PlasticsEurope.

PlasticsEurope (2015) *Factsheet: An Analysis of European Plastics Production, Demand and Waste Data*. Brussels, PlasticsEurope.

Plaxis (2011) *PLAXIS Reference Manual*. Delft, Plaxis.

Rahardjo, H., Li, X.W., Toll, D.G. & Leong, E.C. (2001) Effect of antecedent rainfall on slope stability. *Geotechnical and Geological Engineering*, 19 (3–4), 371–399.

Rahimi, A., Rahardjo, H. & Leong, E.C. (2010) Effect of hydraulic properties of soil on rainfall-induced slope failure. *Engineering Geology*, 114 (3), 135–143.

Rajashree, S.S. & Sitharam, T.G. (2001) Nonlinear finite-element modeling of batter piles under lateral load. *Journal of Geotechnical and Geoenvironmental Engineering*, 127 (7), 604–612.

Rebeiz, K.S. & Craft, A.P. (1995) Plastic waste management in construction: technological and institutional issues. *Resources, Conservation and Recycling*, 15 (3), 245–257.

Rogers, L.E. & Wright, S.G. (1986) *The Effects of Wetting and Drying on the Long-Term Shear Strength Parameters for Compacted Beaumont Clay, Research Report 436-2F*. Center for Transportation Research, University of Texas at Austin.

Saleh, A.A. & Wright, S.G. (1997) *Shear Strength Correlations and Remedial Measure Guidelines for Long-Term Stability of Slopes Constructed of Highly Plastic Clay Soils, Research Report 1435-2F*. Center for Transportation Research, University of Texas at Austin.

Santi, P.M., Elifrits, C.D. & Liljegren, J.A. (2001) Design and installation of horizontal wick drains for landslide stabilization. *Transportation Research Record: Journal of the Transportation Research Board*, 1757 (1), 58–66.

Short, R. & Collins, B.D. (2006) Testing and evaluation of driven plate piles in full-size test slope: New method for stabilizing shallow landslides. In: *TRB 85th Annual Meeting Compendium of Papers CD-ROM, January 22–26, Washington, DC.*

Short, R., Collins, B.D., Bray, J.D. & Sitar, N. (2005) *Testing and Evaluation of Driven Plate Piles in a Full Size Test Slope: A New Method for Stabilizing Shallow Landslides.*

Singh, N., Hui, D., Singh, R., Ahuja, I.P.S., Feo, L. & Fraternali, F. (2016) Recycling of plastic solid waste: a state of art review and future applications. *Composites Part B: Engineering*, Online Version.

Skempton, A.W. (1970) First-time slides in over-consolidated clays. *Geotechnique*, 20 (3), 320–324.

Skempton, A.W. (1977) Slope stability of cuttings in Brown London Clay. In: *Proc. Ninth Int. Conf. on Soil Mechanics and Foundation Engineering, Tokyo*. Vol. 3. pp. 261–270.

Sommers, L., Loehr, J.E. & Bowders, J.J. (2000) Construction methods for slope stabilization with recycled plastic pins. In: *Proceedings of Mid-Continent Transportation Symposium, 15–16 May 2000, Ames, IA*. Ames, IA, Center for Transportation Research and Education, Iowa State University.

Song, B., Xue, L. & Long, F. (2011) Quantitative study on the generation characteristics of domestic plastics waste in Beijing. *Journal of Basic Science and Engineering*, 19 (2), 211–220.

Subramanian, P.M. (2000) Plastics recycling and waste management in the US. *Resources, Conservation and Recycling*, 28, 253–263.

Tarquinio, F. & Pearlman, S.L. (1999) Pin piles for building foundations. In: *Presented at the 7th Annual Great Lakes Geotechnical and Geoenvironmental Conference, 10 May 1999, Kent, OH.*

Thompson, R.C., Swan, S.H., Moore, C.J. & vom Saal, F.S. (2009) Our Plastic Age. *Philosophical Transactions of the Royal Society B: Biological Sciences*, 364, 1973–1976.

Titi, H. & Helwany, S. (2007) *Investigation of Vertical Members to Resist Surficial Slope Instabilities (No. WHRP 07-03)*. Madison, WI, Wisconsin Department of Transportation.

Turner, A.K. & Schuster, R.L. (1996) *Landslides: Investigation and Mitigation. Transportation Research Board Special Report 247*. Washington, DC, National Research Council.

UNEP (2014) *Valuing Plastics: The Business Case for Measuring, Managing and Disclosing Plastic Use in the Consumer Goods Industry*. Nairobi, United Nations Environment Programme.

USDA (1992) *Natural Resources Conservation Service, National Engineering Handbook, Part 650, Engineering Field Handbook, Chapter 18, Soil Bioengineering for Upland Slope Protection and Erosion Reduction*. Washington, DC, United States Department of Agriculture.

USEPA (2015) *Advancing Sustainable Materials Management: 2013 Fact Sheet – Assessing Trends in Material Generation, Recycling and Disposal in the United States*. Washington, DC, United States Environmental Protection Agency.

USGS (2005) *Texas Geologic Map Data*. United States Geological Survey. Available from: https://mrdata.usgs.gov/geology/state/state.php?state=TX [Accessed 25th July 2013].

Vanapalli, S.K., Fredlund, D.G., Pufahl, D.F. & Clifton, A.W. (1996) Model for the prediction of shear strength with respect to soil suction. *Canadian Geotechnical Journal*, 33 (3), 379–392.

Van Genuchten, M.T. (1980) A closed-form equation for predicting the hydraulic conductivity of unsaturated soils. *Soil Science Society of America Journal*, 44 (5), 892–898.

Van Ness, K.E., Nosker, T.J., Renfree, R.W., Sachan, R.D., Lynch, J.K. & Garvey, J.J. (1997) Creep behavior of commercially produced plastic lumber. In: *Technical Papers of the 56th Annual Technical Conference, Society of Plastics Engineers, Atlanta*. Vol. 3. Society of Plastics Engineers. pp. 3128–3135.

Wei, W.B. & Cheng, Y.M. (2009) Strength reduction analysis for slope reinforced with one row of piles. *Computer and Geotechnics*, 36, 1176–1185.

Wright, S.G. (2005) *Evaluation of Soil Shear Strengths for Slope and Retaining Wall Stability Analyses with Emphasis on High Plasticity Clays*. Washington, DC, Federal Highway Administration.

Wright, S.G., Zornberg, J.G. & Aguettant, J.E. (2007) *The Fully Softened Shear Strength of High Plasticity Clays (FHWA/TX-07/0-5202-3)*. University of Texas at Austin.

Yang, S., Ren, X. & Zhang, J. (2011) Study on embedded length of piles for slope reinforced with one row of piles. *Journal of Rock Mechanics and Geotechnical Engineering*, 3 (2), 167–178.

Zhang, Y., Zhao, Y. & Kuang, X. (2006) Analysis on hazardness of plastics packaging wastes in city. *Journal of Guangdong Industry Technical College*, 5 (3), 18–20.

Zornberg, J.G., Kuhn, J. & Wright, S. (2007) *Determination of Field Suction Values, Hydraulic Properties and Shear Strength in High PI Clays, Research Rep. 0-5202-1*. Center for Transportation Research, University of Texas at Austin.

Subject index

Printed and bound by CPI Group (UK) Ltd, Croydon, CR0 4YY

24/10/2024

01778286-0004